# Confronting Climate Uncertainty in Water Resources Planning and Project Design

## *The Decision Tree Framework*

**Advance praise...**

"This is a well-thought out and really useful guidebook for water resources practitioners on sorting through a chaotic accumulation of complex literature and procedures for climate uncertainty analysis, as they pertain to practical methods for the selection and sizing of water management projects under climate uncertainty... This book presents a well-organized and sensible 'bottom-up' approach for achieving valuable insights on the consequences of uncertain climate information for selecting feasible pathways to robust, reliable, and resilient water management solutions."

—**Eugene Stakhiv**, *U.S. Army Corps of Engineers and Johns Hopkins University*

"Nepal is highly vulnerable to climate change and other risks when it comes to water resources and hydropower development. Understanding and managing the risks are crucial for the country's development. The decision tree and its application in the Arun basin in Nepal has demonstrated the value of this method in understanding and managing climate and other risks, including natural disaster risks. We welcome the opportunity to learn more about this method through this book."

—**Durga P. Sangroula**, *Institute of Engineering, Tribhuwan University, Nepal*

"Formal climate change risk analysis is a rigorous approach, but it remains technically challenging, expensive, and often unconvincing in its implementation. This book presents a more tractable and understandable approach that insightfully explores the vulnerability of water projects to the range of changes in climate likely to affect long-term performance."

—**Jay R. Lund**, *Center for Watershed Sciences, University of California-Davis*

"The 'Decision Tree Framework' presented in this book is a valuable tool for guiding and streamlining project design under uncertain climatic and socioeconomic futures... a must-read for researchers and practitioners in the field. A comprehensive review of decision making under uncertainty and insightful discussions are also included."

—**Tingju Zhu**, *International Food Policy Research Institute (IFPRI)*

"Concise and clearly illustrated, the book charts a practical path to robust decisions for planners facing complex choices about large public and private investments. It will become a fixture on the desks of busy practitioners and lead to better outcomes on the ground."

—**Dustin Garrick**, *Department of Political Science and Booth School of Engineering Practice, McMaster University, Ontario, Canada*

"From the most developed watersheds in the world to the least developed, the Decision Tree Framework provides a critical tool for improving climate change risk assessment and decision making. The stepwise approach ensures that climate change risks are considered on every project. It focuses attention and resources on those projects with the greatest risk, providing decision makers with a much clearer picture of the vulnerabilities faced by each project."

—**John Andrew**, *Executive Manager for Climate Change, California Department of Water Resources*

# Confronting Climate Uncertainty in Water Resources Planning and Project Design

## *The Decision Tree Framework*

Patrick A. Ray
Casey M. Brown

**WORLD BANK GROUP**

# Contents

Figures

Table

# Foreword

Water and climate are inseparable. Water is also a naturally variable element, making it difficult to predict and challenging to manage. Today, with the earth warming faster than previously thought, large changes in temperature and precipitation have brought about even more uncertainty concerning the future of our water resources.

Water insecurity—or the lack of the right quantity and quality of water in a given space and time—poses one of the most significant risks for many countries around the world. The threat is most dire in regions that already experience extreme climatic conditions. In arid regions, water scarcity is exposing millions to hunger and other health risks. By 2025, about 1.8 billion people will be living in areas with absolute water scarcity. In other regions, severe floods are responsible for significant human and economic losses. The poorest and most vulnerable are suffering the most.

The primary challenge of achieving water security is our ability to make decisions in the present that sufficiently account for the needs of the future. This becomes particularly important in water projects, especially those that involve investments in long-lived infrastructure that must deliver benefits for many generations to come. Resilient infrastructure will enable countries to respond to floods and droughts, sea level rise, and unpredictable river runoff, and to bring clean and safe water to those currently without access. Moreover, such infrastructure can regulate water flow to prevent disruptions in energy production, agricultural yields, and industrial growth.

At the World Bank, investing in resilience is crucial for achieving our twin goals of reducing poverty and ensuring shared prosperity. Our clients are realizing that careful, climate-smart project design has enormous potential to help them grow sustainably into the 21st century and remain competitive. To meet their demand for transformational, cutting-edge knowledge the World Bank has launched a single Water Global Practice that incubates the best expertise in water on a global scale.

But how do we ensure that our investments are resilient to climate risks? How do we make climate-smart policy choices against a backdrop of uncertainty? So far, the world has relied on global circulation models that provide information on general climate patterns whose projections are downscaled to fit local contexts. However, as useful as this method is for setting the broader context, it does not incorporate the local vulnerabilities to climate change needed to inform investment and policy choices.

To fill this critical knowledge gap, the World Bank Group has partnered with member organizations of the Alliance for Global Water Adaptation, such as the U.S. Army Corps of Engineers, the University of Massachusetts, and the Stockholm International Water Institute, to develop an innovative tool that tackles long-term climate uncertainty in water projects.

I am pleased to present *Confronting Climate Uncertainty in Water Resources Planning and Project Design: The Decision Tree Framework*, a new decision support tool that aims to help project managers and development practitioners to pragmatically assess potential climate risks. The document, developed by the Water Global Practice with the support of our Water Partnership Program (WPP), helps practitioners navigate the maze of existing climate assessment methods and models. The tool first screens for climate vulnerabilities, and a "decision tree" subsequently helps project teams assess and then develop plans to manage climate and other risks. What makes this innovative is its step-by-step design—similar to a tree on which each "branch" builds off the previous one. Further or deeper analysis is performed only as needed, which helps decision makers allocate scarce project resources in a way that is proportional to project needs.

This work represents the first step in a larger effort to apply this framework in projects worldwide, and it is already being piloted by the World Bank in several countries. We hope it will provide a useful framework for making smart choices that generate a climate-resilient future.

*Junaid Kamal Ahmad*
*Senior Director*
*Water Global Practice*
*The World Bank Group*

# Acknowledgments

This work was made possible by the financial contribution of the Water Partnership Program (http://water.worldbank.org/water/wpp) of the Water Global Practice, World Bank Group.

This work was carried out under the direction of Junaid Kamal Ahmad, Senior Director, Water Global Practice, World Bank Group; William Rex, Acting Practice Manager, Water Global Practice; and Julia Bucknall, former Manager of the Water Anchor, the World Bank.

## Authors

Patrick A. Ray (University of Massachusetts Amherst) and Casey M. Brown (University of Massachusetts Amherst).

## Managing and Editing

Luis E. García (World Bank); Diego J. Rodríguez (World Bank); Marcus Wijnen (World Bank).

Technical inputs provided by Mehmet Umit Taner, Johan Grijsen, David Groves, Robert Lempert, Jan Kwakkel, Laura Bonzanigo, and Robert L. Wilby, as well as the comments, suggestions, and guidance received from the participants in the internal and external decision tree discussion workshops, are gratefully acknowledged. Special thanks to the Bank peer reviewers Rikard Liden and Xiaokai Li, who provided valuable guidance and suggestions for this book. Thanks also to the English and publishing editor, Inge Pakulski.

# About the Authors

**Patrick Ray** is research assistant professor in the Department of Civil and Environmental Engineering at the University of Massachusetts Amherst and a member of the Hydrosystems Research Group. His academic work has been supported by numerous grants and awards, including a National Science Foundation Graduate Research Fellowship and a Fulbright Fellowship. While completing a PhD at Tufts University, he was the primary investigator on two studies of water system planning under uncertainty in the Middle East. He has served as an expert consultant for the World Bank, the International Finance Corporation, the United Nations Development Programme, and the World Food Program, among others. In 2010 he served as economist for MWH Global on the Millennium Challenge Corporation's $100 million Zarqa Governorate Wastewater System Expansion and Rehabilitation Project in Jordan.

**Casey Brown** is associate professor in the Department of Civil and Environmental Engineering at the University of Massachusetts Amherst and director of the Hydrosystems Research Group. He has a PhD in environmental engineering science from Harvard University and led the water team at the International Research Institute for Climate and Society at Columbia University. He has a number of awards to his credit, including the Presidential Early Career Award for Science and Engineering, the National Science Foundation CAREER award, and the Huber Research Prize from the American Society of Civil Engineers (ASCE). Dr. Brown is associate editor of *Water Resources Research* and the ASCE *Journal of Water Resources Planning and Management*, chair of the Water Resources Planning under Climate Change Technical Committee of the ASCE Environmental and Water Resources Institute Systems Committee, and past chair of the Water and Society Technical Committee of the American Geophysical Union's Hydrology Section.

# Executive Summary

No methodology has yet been generally accepted for assessing the significance of climate risks relative to all other risks to water resources projects. The need for such a process has recently been elevated in the World Bank. For example, as of December 2013, all International Development Association (IDA) Country Partnership Frameworks must include climate- and disaster-risk considerations in the analysis of the country's development priorities, and, when agreed upon with the country, such considerations must be incorporated into the content of the development programs.

The goal of this book is to outline a pragmatic process for risk assessment of water resources projects that can serve as a decision support tool to assist project planning under uncertainty and that would be useful for the World Bank as well as for other practitioners. The approach adopted here is a robustness-based, bottom-up alternative to previous top-down approaches to climate risk assessment, the quality of which has been contingent on the accuracy of future climate projections derived from general circulation models (GCMs).

Though considerable investment has been made in climate modeling and the downscaling of GCMs with the aim of benefiting decision makers, a recent study by the World Bank's Independent Evaluation Group (IEG) found that "climate models have been more useful for setting context than for informing investment and policy choices" (IEG 2012, 61) and "they often have relatively low value added for many of the applications described" (IEG 2012, 69). The lack of success in the use of climate projections to inform decisions is not due to lack of effort in translating model outputs to be relevant to decision makers. Instead, two fundamental and unavoidable issues limit the utility of these approaches.

The first issue could be classified as a risk *assessment* problem. The uncertainty associated with future climate is largely irreducible in the temporal and spatial scales that are relevant to water resources projects. As a result, climate science–led efforts do not typically reduce the uncertainty of future climate and, in fact, are unlikely to describe the limits of the range of possible climate changes. Perhaps most important, GCMs have the least skill in generating the variables that are most important for water resources projects, such as local hydrologic extremes like floods and drought. Often, the results of a climate change analysis present a wide range of possible future mean climates without providing any insight into climate extremes, and give the sense that they only show the tip of the iceberg for climate uncertainty. As a result, the project planner gains little insight into the potential impact of climate change on the project.

The second issue relates to risk *management*. If climate-related risks are quantified in the process of climate change risk assessment, it remains unclear in most cases whether the effects of changes in climate on a certain water resources project are significant relative to the impacts of changes in other nonclimate factors (such as demographic, technological, land use, and economic changes). Project planners are therefore ill-equipped to incorporate uncertain climate information into a broader (all-uncertainty) assessment of a project's probability of success, and thus to make intelligent modifications to the project design to reduce its vulnerabilities to failure. And if the project planner succeeds in characterizing the relative importance of various risks and system vulnerabilities, the choice remains as to how best to manage those risks to improve system robustness and flexibility. Though a number of analytical tools have been developed, engagement with the tools can be complex and expensive.

In the typical engagement with science, the scientific analysis reduces uncertainty and identifies a likely future, at which point the planner can select the best options for that future. However, given that climate science is not in a position to present a likely future of limited and reasonable range, a different approach to project assessment and decision making is needed. This book puts forth a decision support tool in the form of a "decision tree" to meet this need (see figure ES.1).

The decision tree provides guidance on the application of proven techniques for climate change risk assessment and advanced tools for risk management. Decision scaling, upon which the decision tree's general structure is based, is a bottom-up, robustness-based approach to water system planning that uses a stress test for the identification of system vulnerabilities and simple, direct techniques for the iterative reduction of system vulnerabilities through targeted design modifications. Decision scaling features prominently in the risk-assessment aspects of the decision tree because it is efficient and scientifically

# IDENTIFYING AND MANAGING CLIMATE RISKS

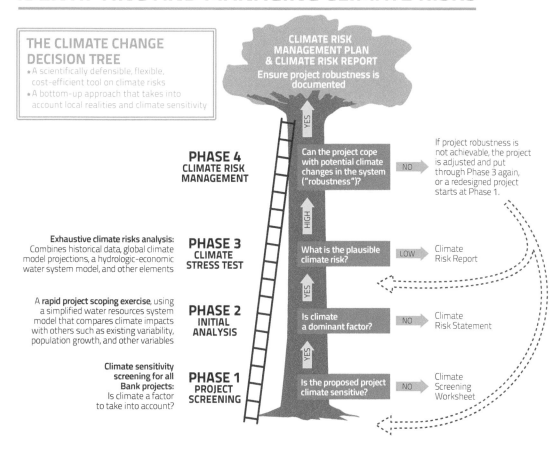

**THE CLIMATE CHANGE DECISION TREE**
- A scientifically defensible, flexible, cost-efficient tool on climate risks
- A bottom-up approach that takes into account local realities and climate sensitivity

**CLIMATE RISK MANAGEMENT PLAN & CLIMATE RISK REPORT**
Ensure project robustness is documented

**PHASE 4**
CLIMATE RISK MANAGEMENT

Can the project cope with potential climate changes in the system ("robustness")? — NO — If project robustness is not achievable, the project is adjusted and put through Phase 3 again, or a redesigned project starts at Phase 1.

HIGH

Exhaustive climate risks analysis: Combines historical data, global climate model projections, a hydrologic-economic water system model, and other elements

**PHASE 3**
CLIMATE STRESS TEST

What is the plausible climate risk? — LOW — Climate Risk Report

YES

A **rapid project scoping exercise**, using a simplified water resources system model that compares climate impacts with others such as existing variability, population growth, and other variables

**PHASE 2**
INITIAL ANALYSIS

Is climate a dominant factor? — NO — Climate Risk Statement

YES

Climate sensitivity screening for all Bank projects: Is climate a factor to take into account?

**PHASE 1**
PROJECT SCREENING

Is the proposed project climate sensitive? — NO — Climate Screening Worksheet

defensible (that is, it does not involve numerous assumptions about the future, nor does it rely on GCMs for direct climate input). Moreover, it makes the best use of available climate change projections, which, though highly uncertain, can still be useful under particular conditions. It creates those conditions by assessing the relative performance and vulnerabilities of alternatives; using that information to describe future scenarios; and then applying the available information regarding local climate trends, climate projections, and historical climate variability to answer specific questions that arise during the decision-making process.

In addition to addressing the fundamental science issues, the decision tree framework was designed with the economic use of human and financial resources in mind. The goal was to develop a tool that would be applicable to all water resources projects, and would also allocate effort to projects in a way that is consistent with each project's potential sensitivity to climate risk. To accomplish this goal, the process was designed to be hierarchical, with different phases of analysis triggered by the findings of the previous phase. The procedure consists of four successive phases—Phase 1: Project Screening; Phase 2: Initial Analysis; Phase 3: Climate Stress Test; and Phase 4: Climate Risk Management. The result is that different categories of projects would be subjected to different types of analysis with effort that is proportionate to the need. The methodology is illustrated using an example of run-of-the-river hydropower development planning in chapter 4.

The particular strength of decision scaling lies in risk assessment. Other tools for decision making under uncertainty, such as robust decision making, stochastic and robust optimization (including real options analysis), dynamic adaptive policy pathways, or information-gap decision theory, enhance the capabilities of the decision tree in risk management. Chapter 5 summarizes a number of the most relevant tools. A selection of hydrologic models that may be incorporated into the decision tree framework is reviewed in appendix A.

A project manager must account for many kinds of uncertainties when seeking approval for funding from a donor or board; most important, the project must be shown to be cost-effective, flexible, and robust. The decision tree described here provides the project manager with a scientifically defensible, repeatable, direct, and clear method for demonstrating the robustness of a project to climate change. At the conclusion of the process, the project manager will be empowered to confidently communicate the method by which the vulnerabilities of the project have been carefully assessed, and how any necessary adjustments improved the project's feasibility and profitability.

## Reference

IEG (Independent Evaluation Group). 2012. *Adapting to Climate Change: Assessing the World Bank Group Experience*. Washington, DC: World Bank.

# Abbreviations

| | |
|---|---|
| AGWA | Alliance for Global Water Adaptation |
| CBA | cost-benefit analysis |
| CLM | Community Land Model |
| CMIP3 | Phase 3 of the Coupled Model Intercomparison Project |
| CMIP5 | Phase 5 of the Coupled Model Intercomparison Project |
| DAPP | Dynamic Adaptive Policy Pathways |
| EIRR | economic internal rate of return |
| GCM | general circulation model |
| GIS | geographic information system |
| GLUE | generalized likelihood uncertainty estimation |
| GWh | gigawatt hour |
| IADB | Inter-American Development Bank |
| IDA | International Development Association (the World Bank's fund for the poorest) |
| IGDT | information gap decision theory |
| IPCC | Intergovernmental Panel on Climate Change |
| $k$-NN | $k$-Nearest Neighbor |
| kWh | kilowatt hour |
| m³/s | cubic meters per second |
| mm | millimeter |
| MW | megawatt |
| mWh | megawatt hour |
| NPV | net present value |

| PDF | probability density function |
| PFT | plant functional type |
| PI | performance indicator |
| PRIM | patient rule induction method |
| PRMS | Precipitation Runoff Modeling System |
| Q | runoff/streamflow |
| RCP | representative concentration pathway |
| RDM | robust decision making |
| RO | (multiobjective) robust optimization |
| ROA | real options analysis |
| SEI | Stockholm Environment Institute |
| SWAT | Soil and Water Assessment Tool |
| UKCIP | United Kingdom Climate Impacts Programme |
| UNDP | United Nations Development Programme |
| USAID | United States Agency for International Development |
| USGS | United States Geological Survey |
| VBA | Visual Basic for Applications (programming language) |
| VIC | Variable Infiltration Capacity (hydrologic) model |
| WARM | Wavelet Autoregressive Model |
| WATBAL | Water Balance (the hydrologic modeling component of WEAP) |
| WEAP | Water Evaluation and Planning system (modeling software) |
| WRI | World Resources Institute |

**CHAPTER 1**

# Introduction

Water infrastructure projects are a significant portion of the World Bank's lending portfolio and a major need for developing countries throughout the world. Many water resources projects have long periods of economic return, with significant uncertainties in the behavior of the natural system as well as that of human factors (technology, population dynamics, economic development, and the like). Traditionally, attempts to quantify and incorporate those uncertainties in planning tools have assumed stationarity of historical trends. For example, the concept of hydrologic stationarity historically has been deemed adequate for water project design with at least partial understanding that hydrologic *response* (due to land use changes) and hydrologic *variability* (due to climate variability and change) are not fully stationary. The uncertainties associated with climate change, however, have led to a reconsideration of whether the water development community is adequately taking into account the uncertainties that characterize the future (Milly et al. 2008). This additional scrutiny of water resources projects is warranted, given the large potential regrets associated with possible stark climate changes in the future. However, the means for conducting this additional assessment is unclear. No general methodology has been accepted for assessing the significance of climate risks relative to all other risks for water resources projects, nor is there an accepted process within the Bank.

The need for an accepted process for climate change risk assessment at the Bank has recently been elevated. Under the IDA17 Replenishment[1] (which added $52 billion to the resources of the International Development Association [IDA], the Bank's fund for the poorest), Special Theme 3 on climate change calls for all IDA Country Partnership Frameworks to incorporate climate- and disaster-risk considerations into the analysis of the country's development challenges and priorities, and, when agreed upon with the country, to incorporate such considerations in the content of the programs and the results framework.

The goal of this book is to outline a pragmatic process for risk assessment of Bank water resources projects that can serve as a decision support framework—a "decision tree"—to assist project planning under uncertainty.

The decision tree described in this book is based on the growing consensus that robustness-based approaches are needed to address uncertainty and its potential impacts on infrastructure planning (Wilby and Dessai 2010). These approaches emphasize assessment of individual projects and their ability to perform well over a wide range of future uncertainty, including climate and other uncertainties (Brown and Wilby 2012; Hallegatte et al. 2012; Prudhomme et al. 2010). Together, they are an alternative to the most prominent kind of approaches to risk assessment, so-called top-down approaches, which assess system response under a limited set of plausible future climate, demographic, and land use conditions, with climate projections downscaled from time series of general circulation models (GCMs). Considerable investment has been made in climate modeling and downscaling with the aim of benefiting decision makers. However, a recent study by the Bank's Independent Evaluation Group (IEG) found that "climate models have been more useful for setting context than for informing investment and policy choices" (IEG 2012, 61) and "they often have relatively low value added for many of the applications described" (IEG 2012, 69).

The lack of success in the use of climate projections to inform decisions is not due to lack of effort in translating model outputs to a form relevant to decision makers. Instead, two fundamental and unavoidable issues limit the utility of these approaches. The first could be classified as a risk *assessment* problem. The uncertainty associated with future climate is largely irreducible in the temporal and spatial scales that are relevant to water resources projects (Stainforth et al. 2007). Climate projections provide limited and often biased explorations of the effects of internal climate variability, especially precipitation variability (Rocheta et al. 2014), with amplified carryover effects for runoff estimates (Fekete et al. 2004). As a result, climate science–led efforts do not typically reduce the uncertainty of future climate in a way that is relevant for water systems planning and, in fact, are unlikely to describe the limits of the range of possible climate changes. Nor are they able to provide probabilistic representations of the uncertainty (Hall 2007).

Because risk is a function of both probability and impact (Dessai and Hulme 2004), the inability of climate projections to probabilistically represent uncertainty is a substantial impediment to the assessment of climate-related risks for proposed water resources projects. Perhaps most important, GCMs have the least skill in generating the variables that are most important for water resources projects, such as local hydrologic variability and extremes (for example, flood and drought). Those extreme events are located at the tails of distributions of climate variables and percentage-wise will change more rapidly than the mean in a changing climate (Dai, Trenberth, and Karl 1998). Often, the results of a climate change analysis present a wide range of possible future mean climates, without any insight into climate extremes, and give the sense that they are only the tip of the iceberg for climate uncertainty. As a result, the project planner gains little insight into the potential impact of climate change on the project.

The second issue relates to risk *management*. As described in chapter 2, some researchers have suggested that the magnitude of the effects of changes in climate on water resources might be small relative to the impact of changes in other variables such as population, technology, and demand, over medium- to long-range periods (for example, Frederick and Major 1997; Lins and Stakhiv 1998); others have suggested that climate change may be the most important factor for long-range development planning (for example, Arnell and Lloyd-Hughes 2014; Rockstrom et al. 2009); and still others have argued that the relative likelihoods of long-range change magnitudes are not quantifiable from our present limited perspective (for example, Allen, Raper, and Mitchell 2001; Lempert et al. 2004). Project planners are therefore ill-equipped to incorporate climate information and all its uncertainties into a broader (all-uncertainty) assessment of a project's probability of success, and thus to make intelligent modifications to the project design to reduce its vulnerabilities to failure. And if the project planner succeeds in characterizing the relative importance of various risks and system vulnerabilities, the choice remains as to how best to manage those risks to improve system robustness and flexibility. Though a number of analytical tools have been developed, engagement with the tools can be complex and expensive.

In the typical engagement with science, the scientific analysis reduces uncertainty and identifies a likely future, as a result of which the planner can select the best options for that future. However, given that climate science is not in a position to present a likely future of limited and reasonable range, a different approach to project assessment and decision making is needed.

The decision tree is designed to address these fundamental issues to provide a path forward for project planners who face decisions potentially affected by climate change uncertainty. In addition to addressing the fundamental science issues described above, the decision tree was also designed so that human and financial resources could be used economically. Water resources projects are quite diverse, including water sector reform, water management, development of hydrometeorological networks, and establishment of new infrastructure, including water supply, sanitation, and hydroelectric facilities. The goal of this work was to develop a tool that would be applicable to all water resources projects, but that would also allocate climate risk assessment effort in a way that is consistent with each project's potential sensitivity to that climate risk. To accomplish this objective, the process was designed to be hierarchical, with different stages or phases of analysis potentially triggered by the findings of the previous phase. The result is that different categories of projects will undergo different types of analysis, with effort that is proportionate to the need.

A project manager must account for uncertainties of many kinds when seeking approval for funding from a donor or board; most important, the project must be shown to be cost-effective and robust. The decision tree described here provides the project manager with a scientifically defensible, repeatable, direct, and clear method for demonstrating the robustness of a project to climate change. At the conclusion of the process, the project manager will be able to confidently communicate the method by which the vulnerabilities of the project were carefully assessed, and how any necessary adjustments improved the project's feasibility and profitability.

Before the decision tree framework is described in chapter 3, chapter 2 provides background on the risks relevant to water systems planning, describes the different approaches to scenario definition in water systems planning, and introduces the decision scaling methodology upon which the general structure of the decision tree framework is based. As explained in chapter 2, decision scaling is a robustness-based, bottom-up approach to the integration of the best current methods of climate risk assessment, which supports simple, direct procedures for risk management. When faced with complex and interconnected system uncertainties, decision scaling can be combined with more advanced risk-management tools, such as robust decision making (Lempert et al. 2006). Chapter 3 describes the decision tree tool, illustrated in figures 3.1 and 3.2, and explains each of the steps and processes that make up the tool. Chapter 4 focuses on a case study of a small hydropower project as an illustration of the decision tree procedure. Chapter 5 describes some of the tools available for decision making under uncertainty and methods available for climate risk management. Concluding thoughts on implementation are presented in chapter 6.

## Note

1. The Final IDA17 Replenishment meeting took place in Moscow, Russia, December 16–17, 2013. Also known as the Bank's fund for the poorest, IDA is the Bank's main instrument for achieving the goals of ending extreme poverty and boosting shared prosperity in the world's poorest countries. The overarching theme of IDA17 was maximizing the development impact per unit of aid—through access to electricity, vaccines, microfinance, basic health services, clean water, sanitation facilities, and so forth.

## References

Allen, M., S. Raper, and J. Mitchell. 2001. "Climate Change – Uncertainty in the IPCC's Third Assessment Report." *Science* 293 (5529): 430–33.

Arnell, N. W., and B. Lloyd-Hughes. 2014. "The Global-Scale Impacts of Climate Change on Water Resources and Flooding under New Climate and Socio-Economic Scenarios." *Climatic Change* 122 (1–2): 127–40.

Brown, C., and R. L. Wilby. 2012. "An Alternate Approach to Assessing Climate Risks." *EOS, Transactions, American Geophysical Union* 92 (41): 401–12.

Dai, A., K. Trenberth, and T. Karl. 1998. "Global Variations in Droughts and Wet Spells: 1900–1995." *Geophysical Research Letters* 25 (17): 3367–70.

Dessai, S., and M. Hulme. 2004. "Does Climate Adaptation Policy Need Probabilities?" *Climate Policy* 4: 107–28.

Fekete, B., C. Vorosmarty, J. Roads, and C. Willmott. 2004. "Uncertainties in Precipitation and Their Impacts on Runoff Estimates." *J. Clim.* 17: 294–304.

Frederick, K. D., and D. C. Major. 1997. "Climate Change and Water Resources." *Climatic Change* 37 (1): 7–23.

Hall, J. 2007. "Probabilistic Climate Scenarios May Misrepresent Uncertainty and Lead to Bad Adaptation Decisions." *Hydrological Processes* 21 (8): 1127–29.

Hallegatte, S., A. Shah, C. Lempert, C. Brown, and S. Gill. 2012. "Investment Decision Making under Deep Uncertainty: Application to Climate Change." Policy Research Working Paper 6193, World Bank, Washington, DC.

IEG (Independent Evaluation Group). 2012. *Adapting to Climate Change: Assessing the World Bank Group Experience.* Washington, DC: World Bank.

Lempert, R. J., D. G. Groves, S. W. Popper, and S. C. Bankes. 2006. "A General, Analytic Method for Generating Robust Strategies and Narrative Scenarios." *Management Science* 52 (4): 514–28.

Lempert, R., N. Nakicenovic, D. Sarewitz, and M. Schlesinger. 2004. "Characterizing Climate-Change Uncertainties for Decision-Makers – An Editorial Essay." *Climatic Change* 65 (1–2): 1–9.

Lins, H., and E. Stakhiv. 1998. "Managing the Nation's Water in a Changing Climate." *Journal of the American Water Resources Association* 34 (6): 1255–64.

Milly, P. C. D., J. Betancourt, M. Falkenmark, R. M. Hirsch, Z. W. Kundzewicz, D. P. Lettenmaier, and R. J. Stouffer. 2008. "Climate Change—Stationarity Is Dead: Whither Water Management?" *Science* 319 (5863): 573–74.

Prudhomme, C., R. L. Wilby, S. Crooks, A. L. Kay, and N. S. Reynard. 2010. "Scenario-Neutral Approach to Climate Change Impact Studies: Application to Flood Risk." *Journal of Hydrology* 390 (3–4): 198–209.

Rocheta, E., M. Sugiyanto, F. Johnson, J. Evans, and A. Sharma. 2014. "How Well Do General Circulation Models Represent Low-Frequency Rainfall Variability?" *Water Resources Research* 50 (3): 2108–23.

Rockstrom, J., Will Steffen, Kevin Noone, Åsa Persson, F. Stuart Chapin, Eric F. Lambin, Timothy M. Lenton, and others. 2009. "A Safe Operating Space for Humanity." *Nature* 461 (September 24): 472–75.

Stainforth, D. A., M. R. Allen, E. R. Tredger, and L. A. Smith. 2007. "Confidence, Uncertainty and Decision-Support Relevance in Climate Predictions." *Philosophical Transactions of the Royal Society A: Mathematical, Physical and Engineering Science* 365 (1857): 2145–61.

Wilby, R. L., and S. Dessai. 2010. "Robust Adaptation to Climate Change." *Weather* 65 (7): 180–85.

**CHAPTER 2**

# Basis for the Decision Tree Framework

## Risk Enumeration

As described by Arnell (1999), in the most general terms, water resources projects are subject to variable and uncertain pressures on the supply side and the demand side. *Supply-side* pressures include climate change (for example, reductions or increases in precipitation and changes in precipitation timing and intensity; increasing temperature, resulting in increased evaporation) as well as environmental degradation (decreasing the fraction of water available for use); infrastructure degradation (for example, the sedimentation of reservoirs, which decreases water storage capacity); and shifting agreements on the transboundary distribution of water resources. Among the *demand-side* pressures are population growth, migration, and concentration (for example, urbanization); shifting agricultural cropping and irrigation patterns; increased environmental demands (for example, low flow requirements in rivers); and economic or technological development or water tariff restructuring that increases or decreases the quantity of water used per capita. Climate change may affect the demand side as well as the supply side, especially as a result of changes in crop evapotranspiration.

Though climate-related pressures receive a large share of attention from governments and the research community (Kundzewicz et al. 2007), the influence of nonclimate factors is significant and in many places even greater than that of climate. Already, many water resources systems are stressed by resource overallocation, political instability, and economic growth (UNESCO et al. 2012). Analogous to climate change projections, for which general circulation model (GCM) ensembles represent only the "lower bound on maximum range of uncertainty" (Stainforth et al. 2007, 2163), projections of nonclimate factors are also highly uncertain. Water use scenarios are "notoriously difficult to make" (Arnell 1999, S33), and human population growth, one driver of water demand, is subject to myriad volatile and poorly understood factors (Cohen 2003).

Guidance on the estimation of the relative impact on water systems performance of climate change and changes in nonclimate factors is available in the form of broad global studies (for example, Alcamo, Floerke, and Maerker 2007; Arnell 1999; Arnell and Lloyd-Hughes 2014; Vorosmarty et al. 2000) and a large number of more geographically targeted case studies. The derivation of rules of thumb regarding the relative importance of climate and nonclimate uncertainties should be approached with caution because the conclusions of the studies are not easily reconcilable. Whereas Vorosmarty et al. (2000) find projected changes in population to be a more significant source of water stress than climate change in the short term (up to 2025), Arnell (1999) argues that the effects of climate change are likely to be more significant in the long term (beyond 2025). Arnell and Lloyd-Hughes (2014) further conclude that uncertainty in projected future impacts of climate change on exposure to water stress and river flooding is dominated by uncertainty in the projected spatial and seasonal pattern of change in climate (especially precipitation), as represented by the available climate models. In studies at the local scale, Lownsbery (2014) finds uncertainties in projected precipitation change, not municipal or industrial development, to be the most significant risk to the Apalachicola-Chattahoochee-Flint (ACF) River Basin; and Frans et al. (2013) determine that growing regional precipitation, not land use change, is the dominant driver of positive runoff trends in the Upper Mississippi River Basin. In contrast, Ray, Kirshen, and Watkins (2012) observe demographic change projections to have a greater impact on the future water balance of Amman, Jordan, than climate change projections. One statistical tool that might help in the partitioning and ranking of the climate and nonclimate uncertainties facing water systems is the analysis of variance,[1] similar to that presented by Hawkins and Sutton (2009) for climate models, and applied to the ACF River Basin by Lownsbery (2014).

In brief, the relative influence of climate and nonclimate factors is particular to each project context, and must be attended to in each unique case. When evaluating the relative importance of climate and nonclimate factors, issues that deserve particular attention are initial conditions (whether water supplies are already stressed), recent local climate and demographic trends, and project lifetime, with longer-lived projects likely to experience greater climate-related pressure.

The baseline expectation is that a proposed water project would perform satisfactorily with a continuation of current demand- and supply-side conditions. Continuation of such conditions, however, cannot be assumed. When evaluating the robustness of a water project to change, it is important to estimate, as specifically as possible, reasonable ranges of shift in all relevant supply- and demand-side factors. Clearly, not all factors are relevant to every water project, and furthermore, though changes in a certain factor might have an effect on system performance, it may be that no reliable local projections for future values of that factor can be established. Effort should therefore be expended according to the relative significance to system performance of shifts in the various factors, as well as the limitations posed by the available data. Once approximate ranges of future values of the relevant change factors have been estimated, the process of factor-specific and cumulative risk assessment can begin.

## Alternative Approaches to Scenario Definition

Schwartz's *The Art of the Long View* (1996) popularized the business-management (and business-growth) concepts of scenario planning. Internally consistent scenarios of the type endorsed by scenario planning are the basis for most approaches to climate change risk assessment and adaptation planning. Scenarios have narrative power to open the imagination to future worlds. The starting point for imagination of future climates has typically been the output of GCMs, which provide the most advanced, model-based sources of scientific information about future climate that are generally available. The next section describes this approach. The use of preexisting climate scenarios is often convenient because of the ready availability of output from GCMs; however, such scenarios are ill-equipped to illuminate a clear path for long-range water system planning because of the issues described in chapter 1. The "Ex Post Scenario Definition" section of this chapter therefore explains an alternative basis for risk analysis in water system planning that is the backbone of decision scaling and other similar approaches: to first subject the proposed project to a scenario-independent climate stress test, and then (ex post) allow the vulnerabilities of the system to define scenarios of interest.

### Ex Ante Scenario Definition

Traditional decision analysis in water systems planning has used scenarios generated from internally consistent storylines of future global development, downscaled for local or regional applications (for example, Christensen and Lettenmaier 2007; Fowler and Kilsby 2007; Lettenmaier et al. 1999; Minville et al. 2009; Ray, Kirshen, and Watkins 2012; Sun et al. 2008). In this approach to scenario development, GCMs provide estimates of future regional (a typical GCM grid is approximately 250 to 600 kilometers to a side) climate conditions (for example, temperature and precipitation) in the context of a variety of possible global economic, technological, and social trends. Because they link climate and demographics, these storylines can provide guidance for projections of both future water availability and future water consumption, although their use is typically limited to projections of future carbon emissions.

Most studies making use of internally consistent storylines reference either some version of the Intergovernmental Panel on Climate Change's (IPCC's) "Special Report on Emissions Scenarios" (Davidson and Metz 2000), which defines four narrative storylines (A1, A2, B1, and B2 to represent different demographic, social, economic, technological, and environmental developments that diverge in increasingly irreversible ways), or the recently updated "representative concentration pathways" (RCP) scenarios (Moss et al. 2010), which define a new set of radiative forcings[2] (RCP8.5, RCP6.0, RCP4.5, and RCP2.6).

The output of a number of GCMs is available for each projected state of the world in the form of future climate time series. These future climate time series can be obtained before any examination of the project under consideration begins, and as such they define an ex ante, bounded (limited, and possibly ill-selected) set of scenarios over which the performance of the project can be tested. Water systems models using ex ante scenarios, also referred to as the "scenario-led" approach, test the performance of the system using a sample of the ex ante futures described by the storylines. Studies of this type translate time series of climate parameters (for example, average daily temperature and precipitation) from the downscaled GCMs for a particular internally consistent scenario (RCP8.5, for instance) into projections of future streamflow using a hydrologic model (examples from flood risk management include Cameron, Beven, and Naden 2000; Loukas, Vasiliades, and Dalezios 2002; Muzik 2002; Prudhomme, Jakob, and Svensson 2003; Willems and Vrac 2011).

Robustness-based approaches to water systems planning make the strategically critical leap of emphasizing preparedness for a range of possible futures. The importance of this ideological shift is not to be understated. However, most planning exercises still base the projections of future climate on downscaled GCMs, tying the relative severity of climate projections to demographic, economic, and technological trends as described by IPCC storylines.

Top-down frameworks using ex ante climate scenarios can help quantify the relative contribution of different components to overall uncertainty for extremes such as low flows (for example, Wilby and Harris 2006). Moreover, very high resolution regional climate models are now being used to investigate the sensitivity of extreme precipitation to temperature forcing (for example, Kendon et al. 2014). In other words, climate models and downscaling methods can be usefully deployed to strengthen the understanding of the physical processes or critical thresholds that drive hydrological extremes.

However, top-down climate assessments rely heavily on GCM outputs for delineation of the ex ante scenarios describing local and regional climate impacts. Most top-down approaches begin with a small selection of all available GCM output, which, because each GCM will represent the particular climate dynamics of relevance to the project with varying skill, greatly increases the risk that the wrong (unskilled) GCM subset is used. Even if the output from all available GCMs is used, the GCMs represent only a subset of all possible climate futures. This is evidenced in the increase in uncertainty in the range of climate forecasts of the Coupled Model Intercomparison Project (CMIP) from phase 3 to phase 5—the phase 3 ensemble was an underestimate of the phase 5 uncertainty, which may be an underestimate of the uncertainty in the next CMIP phase. As a result, top-down methods do not sample from the full range of climate futures, and may sample from the wrong range entirely. The process of downscaling GCMs results in a cascade of uncertainty (Wilby and Dessai 2010). Furthermore, all models have similar resolution and must put parameters around the same processes (Tebaldi and Knutti 2007). Uncertainties that are related to the underlying science will be the same in different models.

Top-down climate change analyses present a wide range of possible *mean* future climate conditions, but they do not adequately describe the *range* of potential future conditions more generally (Stainforth et al. 2007). In addition, top-down analyses provide limited insight into the changes to climate drivers (such as monsoon patterns and atmospheric rivers) and climate extremes (Olsen and Gilroy 2012). They provide the least information for the variables that are most important for water resources projects, such as hydrologic variability and extremes (for example, floods and droughts). Therefore, deriving probability distributions of these events from an ex ante ensemble of GCMs is fraught with problems. Given the essential role of likelihood concepts in risk assessment (in which risk is a function of impact and probability of that impact), top-down methods tend *not* to provide the insights needed for water resources system planning.

A subtle point in this discussion deserves emphasis: scenario-led studies tend by necessity to select a small subset of all possible futures because of the large computational effort required for generating usable time series from GCM projections for the spatial and temporal context required, in addition to generating internally consistent combinations of uncertain parameters and repeatedly testing the performance of the system. This approach is often prohibitively inefficient. Without targeted information about the vulnerabilities of the system, great effort might be expended in the development and evaluation of ex ante scenarios that offer no further information regarding the system design because they fail to exert stress on the system, or because they do not provide information about which future is more likely nor delimit the range of climate change and variability that might be experienced.

It should be noted that downscaled GCM output is not the only source of climate information in the scenario-led approach. In fact, it is simply an updated input for traditional water resources analysis that had been based on historical time series. Scenarios have been developed through the perturbation and stochastic resampling of historical flow patterns using Markov Chain bootstrap techniques (Lall and Sharma 1996; Sharma, Tarboton, and Lall 1997; Yates et al. 2003) or the assignment of a "change factor" to historical temperature and precipitation values based on climate shifts identified by GCM output (Hay, Wilby, and Leavesley 2000). Use of these other sources of climate information could result in improved accuracy relative to downscaled GCM output (and could better represent local drought and flood risks, particularly), but it does not solve the problem of the inefficiency of scenario development and testing.

Generally, successful approaches to scenario-led planning require many thousands (if not hundreds of thousands or millions) of runs of sophisticated scenario-generating algorithms to trace out the decision space. Successful approaches also tend to require heavy involvement from high-level experts capable of guiding the process.

## Ex Post Scenario Definition

In contrast to ex ante approaches, scenarios can be generated by parametrically (Ben-Haim 2006) or stochastically (Brown et al. 2011; Prudhomme et al. 2010) varying the climate (and other) data to identify vulnerabilities in water system performance, and elaborating scenarios ex post according to the vulnerabilities of the project or the opportunities it presents. Water system models using ex post scenarios test the performance of the system across a very wide range of potential futures (climate and nonclimate permutations), beyond the scope of the futures suggested by the IPCC narratives and associated GCM models. Scenarios are thereby defined as those futures in which the system struggles or fails. Scenarios defined ex post are meant to identify suboptimal system performance, and are less likely to underestimate vulnerabilities. In addition, the ex post definition of scenarios may facilitate the assignment of relative or subjective probabilities to the scenarios.

As discussed in greater depth later in the book, ex post approaches, although they may involve nearly as many model simulations as ex ante approaches, reduce the computational effort by saving scenario definitions that describe only those futures in which the project has demonstrated vulnerabilities. By taking an ex post approach to scenario development, the additional level of effort expended by ex ante approaches on the presimulation development of internally consistent scenarios is not required.

An additional benefit of ex post scenario definition, if scenario definition is necessary at all, is that the considerable uncertainty associated with climate change projections does not enter the analysis until the stage of assigning probabilities. By withholding the use of the projections until the end of the modeling process, rather than inserting them at the beginning, the problem of propagation of uncertainty throughout the analysis (Carter et al. 2007; Jones 2000) is ameliorated. The identification of "narrative scenarios" or clusters of scenarios that are descriptive of futures of concern (to which, for example, the system under consideration is vulnerable) also becomes possible (Groves and Lempert 2007). When only a few uncertainties are considered, such as climate sequences differentiated by average precipitation and temperature, such scenario clusters can be identified visually. When analyses consider more dimensions of uncertainty, scenario-discovery algorithms can be useful (Lempert et al. 2006); these algorithms identify and display the most important combinations of uncertainties affecting system performance. Such scenarios can be useful in communicating information about system vulnerabilities to decision makers (Parker et al. 2013).

## BOX 2.1

# Ex Ante versus Ex Post Scenario Development

Ex ante scenario development is the generation of scenarios from internally consistent storylines of future global conditions (including, but not limited to, demographic, economic, land use, and climate conditions, or skipping straight to radiative forcing) prior to evaluation of the climate-related vulnerabilities of (or opportunities for) the project under investigation. Water system models using ex ante scenarios, also referred to as the "scenario-led" approach, test the performance of the system across a sample of futures described by the storylines. Ex ante scenarios tend to refer to Intergovernmental Panel on Climate Change (IPCC) storylines, and water system evaluations using ex ante scenarios tend to take climate information directly from general circulation models (GCMs).

Ex post scenario development is the generation of scenarios by parametrically or stochastically varying the climate (and other) data to identify vulnerabilities in water system performance, and elaborating scenarios according to the vulnerabilities of the project or the opportunities it presents. Scenarios defined ex post aim to identify suboptimal system performance (in whatever future state that might occur), and are less likely to underestimate vulnerabilities. In addition, the ex post definition of scenarios may facilitate the assignment of relative or subjective probabilities to the scenarios. Because GCM projections are not entered into ex post scenarios until the end of the analysis, ex post scenarios are less susceptible to the well-documented shortcomings of GCMs.

*Exploring Vulnerabilities: Weather Generators and the Climate Stress Test*

To define ex post scenarios, the system's performance needs to be explored. Internally consistent coupled projections of climate and nonclimate factors are typically used to conduct this exploration. Alternatively, a stress test can be used to efficiently explore the performance of alternatives and identify weaknesses. The stress test is a process by which an option is exposed to a variety of plausible climate and nonclimate changes, modifying means and other aspects of variability, to identify vulnerabilities. The analytical engine of the climate aspects of the stress test is an update to traditional weather generators.

Weather generators are computer algorithms that produce long series of synthetic daily weather data. The parameters of the model are conditioned on existing meteorological records to ensure that the characteristics of historical weather emerge in the daily stochastic process. Weather generators are a common tool for extending meteorological records (Richardson 1985); supplementing weather data in a region of data scarcity (Hutchinson 1995); disaggregating seasonal hydroclimatic forecasts (Wilks 2002); and downscaling coarse, long-term climate projections to fine-resolution, daily weather for impact studies (Kilsby et al. 2007; Wilks 1992). A major benefit afforded by most weather generators is their utility in performing climate-sensitivity analyses (Wilks and Wilby 1999). Several studies have used weather generators to systematically test the climate sensitivity of impact models, particularly in the agricultural sector (Confalonieri 2012; Dubrovsky, Zalud, and Stastna 2000; Mearns, Rosenzweig, and Goldberg 1996; Riha, Wilks, and Simoens 1996; Semenov and Porter 1995). These sensitivity studies systematically change parameters in the model to produce new sequences of weather variables (precipitation, for instance) that exhibit a wide range of change in their characteristics (such as average amount, frequency, intensity, and duration).

In the context of a climate stress test, a stochastic weather generator can be built for a region of interest and used to generate several scenarios of daily climate within which a water resources system can be tested. The flexibility of stochastic weather generators enables many climate permutations to be generated, each of which can exhibit a different type of climate alteration that the analyst may be interested in. Note that the permutations created by the weather generator are not dependent on any climate projections at this point in the analysis, thereby allowing a wide range of possible future climates to be generated while avoiding biases propagated from the projections. However, the particular permutations generated can be informed by available projections to ensure that they more than encompass the range of GCM projections. The stochastically generated climate permutations form the foundation of the climate vulnerability assessment. Many different weather generators are mentioned in the scientific literature for use in the climate stress test; a unique model that incorporates low-frequency variability was developed for the case study presented in chapter 4.

### Paleoclimatology Data

One aspect of risk estimation that has traditionally received less attention but is becoming more important to the understanding of likely future climate scenarios is the use of paleoclimatology data to inform expectations for the future. Paleoclimatology data make available much extended data sets (500 or more years), which would be important to inform the risks to which the water system could be exposed. These data are based on observations from the history of a given location, and not on projections of what recently unprecedented conditions might be realized in the future. Often, unfortunately, paleoclimatology data are only available for specific variables and at coarse temporal resolution (annual or decadal). Risks associated with a nonstationary climate have been presented as deviations from observations of the past 100 years or

BOX 2.2
## Bottom-Up, Climate-Informed Decision Making

The term "bottom-up," as adopted in this book, refers to both the emphasis on system vulnerabilities (through ex post scenario development, including late-stage incorporation of climate projection information from general circulation models), and the active involvement of stakeholders in the decision-making process.

so of record; however, natural climate cycles leading to extreme flooding and drought that repeat on periods greater than a single century are likely to provide much better information about the risks faced in the economic lifetime of long-lived water infrastructure (e.g., dams, canals).

Paleoclimatology data have featured prominently in climate risk assessments of the Colorado River (for example, Meko and Woodhouse 2011; Prairie and Rajagopalan 2007) and in decision scaling applications for the Great Lakes (for example, Brown et al. 2011; Moody and Brown 2012).

### Stakeholder Expertise: Top-Down and Bottom-Up
In climate change planning, other studies (for example, Prudhomme et al. 2010) have defined "top-down" to refer to the scenario-led use of downscaled GCMs to describe future regional climate, inputting the scenarios into impact models, and then prescribing adaptations. By contrast, "bottom-up" has been described as ex post scenario development (that is, *not* beginning with ex ante scenarios encompassed in downscaled GCM output). Bottom-up in this sense does not solely refer to the origination of project motivations, design parameters, and performance thresholds with stakeholders. The use of the term "bottom-up" as adopted by the decision tree framework encompasses both aspects—decision-making pathways (beginning with stakeholders) and climatological (ex post climate scenario definition).

## Background on Decision Scaling

Decision scaling (also referred to as climate informed decision analysis) is an approach to the integration of the best current methods for climate risk assessment and robust decision analysis with simple procedures for risk management. Decision scaling, as outlined in figure 2.1, is a robustness-based approach to water system planning that makes use of a stress test for the identification of system vulnerabilities, and simple, direct techniques for the

**FIGURE 2.1 Schematic Comparison of Decision Scaling with Traditional Approach to Climate Change Risk Assessment**

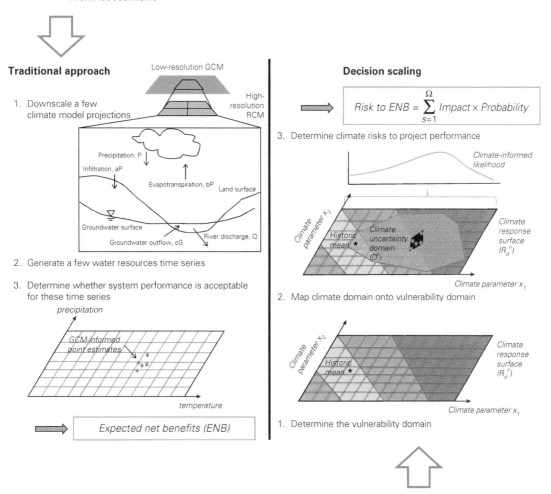

*Note:* GCM = general circulation model; RCM = regional climate model.

iterative reduction of system vulnerabilities through targeted design modifications. The decision scaling methodology has been presented in a number of publications (for example, Brown 2010, 2013; Brown et al. 2011; Brown et al. 2012; Hallegatte et al. 2012). Other tools are discussed in the "Risk Assessment Tools" section of chapter 5.

Decision scaling features prominently in both the risk-assessment aspects and general structure of the decision tree (see chapter 3) because it is efficient and scientifically defensible (that is, it does not involve numerous ex ante assumptions about the future, nor rely on GCMs for direct climate input).

Moreover, it makes the best use of available climate information, which is typically highly inexact but may still be useful under certain conditions. Decision scaling creates those conditions by assessing the relative performance and vulnerabilities of alternative system designs, using that information to describe ex post scenarios, then applying the available climate information to assist in decision making.

Because various sources of climate information can be applied without rerunning the model, decision scaling makes use of all sources of climate information (for example, frequency analysis of GCM output, historical data, paleoclimatology data, stochastically generated climate simulations, and expert judgment of scientists and stakeholders) to inform the likelihoods of different types of climate change. When climate information is deemed fairly reliable and projections are consistent, model-based probabilistic estimates of risk can be made and risk-weighted decision making can be used. In the event that projections based on the various sources are contradictory, not relevant, or not credible, the process enables the identification of climate sensitivities and provides a framework for addressing potential hazards through robustness approaches (see chapter 5 for examples).

Decision scaling is also bottom-up in its development of decision-making pathways. The first step is stakeholder consultation for identification and characterization of historical system performance, desired future performance thresholds, and vulnerabilities to change. Whereas standard decision analysis requires well-characterized uncertainties, decision scaling was developed to handle poorly characterized uncertainties and make the best use of available information.

By serving as a climate stress test designed to evaluate the response of the system to a wide range of plausible climates, beyond the range of the IPCC GCMs but with guidance from those GCMs, decision scaling determines whether the time- and effort-intensive process of GCM downscaling is likely to be beneficial. The resulting climate response function provides insight into the expected performance of the system in an uncertain future. Decision scaling does not include an explicit framework for risk management, although it contributes many of the informational elements required of a decision tool. For decision scaling to be used as a decision analysis tool, it would need to trade off the benefits and costs of mitigating strategies, either by way of an optimization algorithm or by Monte Carlo simulation. Other tools, such as robust decision making, specialize in risk and vulnerability management and have the potential to greatly expand the risk-management capabilities of the decision analysis process.

A decision scaling analysis can be accomplished in a couple of months if all modeling tools are preexisting and available (at a cost below $100,000), or may take one to two years if all required models must first be developed (at a cost of up to $200,000). For comparisons of level of effort and cost with other methods, see Hallegatte et al. (2012).

### Applications of the Decision Scaling Methodology

Examples of applications of the decision scaling methodology are available for the Niger Basin (Brown 2010) and the Upper Great Lakes in North America (Brown et al. 2011; Moody and Brown 2012, 2013). The application to the Upper Great Lakes is instructive with regard to the need for additional measures to address climate risks once they have been identified. An adaptive management plan has been advocated for the Great Lakes and is currently in development by the International Joint Commission (the binational governing body for the Great Lakes). Given the potential for (and indeed likelihood of) faulty assumptions and unforeseen surprises that threaten the success of the Lake Superior regulation plan, and given the significant magnitude of negative consequences to a large portion of North America in the event of a failure of the regulation plan, contingencies were established for residual risks and for unlikely but catastrophic extreme events.

Decision scaling assists in the development of adaptive management plans that establish contingencies in proportion to the negative consequences of system failure. To sustain monitoring and provide mechanisms for use of the collected data in decision making, an institutional framework for the adaptive management process is required. In other cases, alternative approaches to addressing climate risk may be warranted or more appropriate, as discussed in Phase 4 of the decision tree framework (described in chapter 3).

## Notes

1. Analysis of variance is a statistical method used to test differences between two or more means; inferences are made about the means by analyzing variance.
2. Radiative forcing, as defined by Moss et al. (2010), is "the change in the balance between incoming and outgoing radiation to the atmosphere caused by changes in atmospheric constituents, such as carbon dioxide."

# References

Alcamo, J., M. Floerke, and M. Maerker. 2007. "Future Long-Term Changes in Global Water Resources Driven by Socio-Economic and Climatic Changes." *Hydrological Sciences Journal (Journal Des Sciences Hydrologiques)* 52 (2): 247–75.

Arnell, N. 1999. "Climate Change and Global Water Resources." *Global Environmental Change—Human and Policy Dimensions* 9 (1): S31–S49.

——, and B. Lloyd-Hughes. 2014. "The Global-Scale Impacts of Climate Change on Water Resources and Flooding under New Climate and Socio-Economic Scenarios." *Climatic Change* 122 (1–2): 127–40.

Ben-Haim, Y. 2006. *Info-Gap Decision Theory: Decisions under Severe Uncertainty*, 2nd ed. London: Academic Press.

Brown, C. 2010. "Decision-Scaling for Robust Planning and Policy under Climate Uncertainty." World Resources Report, Washington, DC.

——. 2013. "Climate Risk Assessment of the Coralville Reservoir: Demonstration of the Decision-Scaling Methodology." Unpublished.

——, Y. Ghile, M. Laverty, and K. Li. 2012. "Decision Scaling: Linking Bottom-Up Vulnerability Analysis with Climate Projections in the Water Sector." *Water Resources Research* 48 (9): W09537.

Brown, C., W. Werick, W. Leger, and D. Fay. 2011. "A Decision-Analytic Approach to Managing Climate Risks: Application to the Upper Great Lakes." *Journal of the American Water Resources Association* 47 (3): 524–34.

Cameron, D., K. Beven, and P. Naden. 2000. "Flood Frequency Estimation by Continuous Simulation under Climate Change (with Uncertainty)." *Hydrology and Earth System Sciences* 4: 393–405.

Carter, T. R., R. N. Jones, X. Lu, S. Bhadwal, C. Conde, L. O. Mearns, B. C. O'Neill, M. D. A. Rounsevell, and M. B. Zurek. 2007. "New Assessment Methods and the Characterisation of Future Conditions." In *Climate Change 2007: Impacts, Adaptation and Vulnerability. Contribution of the Working Group II to the Fourth Assessment Report of the Intergovernmental Panel on Climate Change*, edited by M. L. Parry, O. F. Canziani, J. P. Palutikof, P. J. van der Linden, and C. E. Hanson, 133–171. Cambridge, U.K.: Cambridge University Press.

Christensen, N. S., and D. P. Lettenmaier. 2007. "A Multimodel Ensemble Approach to Assessment of Climate Change Impacts on the Hydrology and Water Resources of the Colorado River Basin." *Hydrology and Earth System Sciences* 11: 1417–34.

Cohen, J. 2003. "Human Population: The Next Half Century." *Science* 302 (5648): 1172–75.

Confalonieri, R. 2012. "Combining a Weather Generator and a Standard Sensitivity Analysis Method to Quantify the Relevance of Weather Variables on Agrometeorological Models Outputs." *Theoretical and Applied Climatology* 108 (1–2): 19–30.

Davidson, O., and B. Metz. 2000. *IPCC Special Report—Emissions Scenarios: Summary for Policy Makers*. Intergovernmental Panel on Climate Change.

Dubrovsky, M., Z. Zalud, and M. Stastna. 2000. "Sensitivity of Ceres-Maize Yields to Statistical Structure of Daily Weather Variables." *Climatic Change* 46 (4): 447–72.

Fowler, H. J., and C. G. Kilsby. 2007. "Using Regional Climate Model Data to Simulate Historical and Future River Flows in Northwest England." *Climatic Change* 80 (3–4): 337–67.

Frans, C., E. Istanbulluoglu, V. Mishra, F. Munoz-Arriola, and D. P. Lettenmaier. 2013. "Are Climatic or Land Cover Changes the Dominant Cause of Runoff Trends in the Upper Mississippi River Basin?" *Geophysical Research Letters* 40 (6): 1104–10.

Groves, D. G., and R. J. Lempert. 2007. "A New Analytic Method for Finding Policy-Relevant Scenarios." *Global Environmental Change—Human and Policy Dimensions* 17 (1): 73–85.

Hallegatte, S., A. Shah, C. Lempert, C. Brown, and S. Gill. 2012. "Investment Decision Making under Deep Uncertainty: Application to Climate Change." Policy Research Working Paper 6193, World Bank, Washington, DC.

Hawkins, E., and R. Sutton. 2009. "The Potential to Narrow Uncertainty in Regional Climate Predictions." *Bulletin of the American Meteorological Society* 90 (8): 1095–107.

Hay, L., R. Wilby, and G. Leavesley. 2000. "A Comparison of Delta Change and Downscaled GCM Scenarios for Three Mountainous Basins in the United States." *Journal of the American Water Resources Association* 36 (2): 387–97.

Hutchinson, M. F. 1995. "Stochastic Space-Time Models from Ground-Based Data." *Agricultural and Forest Meteorology* 73 (3–4): 237–64.

Jones, R. 2000. "Managing Uncertainty in Climate Change Projections—Issues for Impact Assessment: An Editorial Comment." *Climatic Change* 45: 403–19.

Kendon, S., N. M. Robert, H. J. Fowler, M. J. Roberts, S. C. Chan, and C. A. Senior. 2014. "Heavier Summer Downpours with Climate Change Revealed by Weather Forecast Resolution Model." *Nature Climate Change* 4: 570–76.

Kilsby, C. G., P. D. Jones, A. Burton, A. C. Ford, H. J. Fowler, C. Harpham, P. James, A. Smith, and R. L. Wilby. 2007. "A Daily Weather Generator for Use in Climate Change Studies." *Environmental Modeling and Software* 22 (12): 1705–19.

Kundzewicz, Z. W., L. J. Mata, N. W. Arnell, P. Döll, P. Kabat, B. Jiménez, K. A. Miller, T. Oki, Z. Sen, and I. A. Shiklomanov. 2007. "Freshwater Resources and Their Management." In *Climate Change 2007: Impacts, Adaptation and Vulnerability. Contribution of Working Group II to the Fourth Assessment Report of the Intergovernmental Panel on Climate Change*, edited by M. L. Parry, O. F. Canziani, J. P. Palutikof, P. J. van der Linden, and C. E. Hanson, 173–210. Cambridge, UK: Cambridge University Press.

Lall, U., and A. Sharma. 1996. "A Nearest Neighbor Bootstrap for Resampling Hydrologic Time Series." *Water Resources Research* 32 (3): 679–93.

Lempert, R. J., D. G. Groves, S. W. Popper, and S. C. Bankes. 2006. "A General, Analytic Method for Generating Robust Strategies and Narrative Scenarios." *Management Science* 52 (4): 514–28.

Lettenmaier, D., A. Wood, R. Palmer, E. Wood, and E. Stakhiv. 1999. "Water Resources Implications of Global Warming: A US Regional Perspective." *Climatic Change* 43 (3): 537–79.

Lownsbery, K. E. 2014. "Quantifying the Impacts of Future Uncertainties on the Apalachicola-Chattahoochee-Flint Basin." Master's thesis, University of Massachusetts Amherst.

Loukas, A., L. Vasiliades, and N. Dalezios. 2002. "Potential Climate Change Impacts on Flood Producing Mechanisms in Southern British Columbia, Canada Using the CGCMA1 Simulation Results." *Journal of Hydrology* 259 (1–4): 163–88.

Mearns, L. O., C. Rosenzweig, and R. Goldberg. 1996. "The Effect of Changes in Daily and Interannual Climatic Variability on Ceres-Wheat: A Sensitivity Study." *Climatic Change* 32 (3): 257–92.

Meko, D. M., and C. A. Woodhouse. 2011. "Application of Streamflow Reconstruction to Water Resources Management." *Developments in Paleoenvironmental Research* 11: 231–61.

Minville, M., F. Brissette, S. Krau, and R. Leconte. 2009. "Adaptation to Climate Change in the Management of a Canadian Water-Resources System Exploited for Hydropower." *Water Resources Management* 23 (14): 2965–86.

Moody, P., and C. Brown. 2012. "Modeling Stakeholder-Defined Climate Risk on the Upper Great Lakes." *Water Resources Research* 48 (10): W10524.

——. 2013. "Robustness Indicators for Evaluation under Climate Change: Application to the Upper Great Lakes." *Water Resources Research* 49 (6): 3576–88.

Moss, R. H., J. A. Edmonds, K. A. Hibbard, R. M. Manning, S. K. Rose, D. P. van Vuuren, T. R. Carter, and others. 2010. "The Next Generation of Scenarios for Climate Change Research and Assessment." *Nature* 463 (7282): 747–56.

Muzik, I. 2002. "A First-Order Analysis of the Climate Change Effect on Flood Frequencies in a Subalpine Watershed by Means of a Hydrological Rainfall-Runoff Model." *Journal of Hydrology* 267 (1–2): 65–73.

Olsen, J. R., and K. Gilroy. 2012. "Risk Informed Decision-Making in a Changing Climate." Prepared for 3rd International Interdisciplinary Conference on Predictions for Hydrology, Ecology, and Water Resources Management, Vienna, Austria, September 24–27.

Parker, A. M., S. Srinivasan, R. J. Lempert, and S. Berry. 2013. "Evaluating Simulation-Derived Scenarios for Effective Decision Support." *Technological Forecasting and Social Change* 91 (February): 64–77.

Prairie, J. R., and B. Rajagopalan. 2007. "A Basin Wide Stochastic Salinity Model." *Journal of Hydrology* 344 (1–2): 43–54.

Prudhomme, C., D. Jakob, and C. Svensson. 2003. "Uncertainty and Climate Change Impact on the Flood Regime of Small UK Catchments." *Journal of Hydrology* 277 (1–2): 1–23.

Prudhomme, C., R. L. Wilby, S. Crooks, A. L. Kay, and N. S. Reynard. 2010. "Scenario-Neutral Approach to Climate Change Impact Studies: Application to Flood Risk." *Journal of Hydrology* 390 (3–4): 198–209.

Ray, P. A., P. H. Kirshen, and D. W. Watkins Jr. 2012. "Staged Climate Change Adaptation Planning for Water Supply in Amman, Jordan." *Journal of Water Resources Planning and Management* 138 (5): 403–11.

Richardson, C. W. 1985. "Weather Simulation for Crop Management Models." *Transactions of the American Society of Agricultural Engineers* 28 (5): 1602–6.

Riha, S. J., D. S. Wilks, and P. Simoens. 1996. "Impact of Temperature and Precipitation Variability on Crop Model Predictions." *Climatic Change* 32 (3): 293–311.

Schwartz, P. 1996. *The Art of the Long View*. New York: Doubleday.

Semenov, M. A., and J. R. Porter. 1995. "Climatic Variability and the Modeling of Crop Yields." *Agricultural and Forest Meteorology* 73 (3–4): 265–83.

Sharma, A., D. Tarboton, and U. Lall. 1997. "Streamflow Simulation: A Nonparametric Approach." *Water Resources Research* 33 (2): 291–308.

Stainforth, D. A., T. E. Downing, R. Washington, A. Lopez, and M. New. 2007. "Issues in the Interpretation of Climate Model Ensembles to Inform Decisions." *Philosophical Transactions of the Royal Society A: Mathematical Physical and Engineering Sciences* 365 (1857): 2163–77.

Sun, G., S. G. McNulty, J. A. M. Myers, and E. C. Cohen. 2008. "Impacts of Multiple Stresses on Water Demand and Supply across the Southeastern United States." *Journal of the American Water Resources Association* 44 (6): 1441–57.

Tebaldi, C., and R. Knutti. 2007. "The Use of the Multi-Model Ensemble in Probabilistic Climate Projections." *Philosophical Transactions of the Royal Society A: Mathematical, Physical and Engineering Sciences* 365 (1857): 2053–75.

UNESCO, World Water Assessment Programme, and UN Water. 2012. *Managing Water under Uncertainty and Risk: United Nations World Water Development Report*, 4th edition. Paris: UNESCO.

Vorosmarty, C., P. Green, J. Salisbury, and R. Lammers. 2000. "Global Water Resources: Vulnerability from Climate Change and Population Growth." *Science* 289 (5477): 284–88.

Wilby, R. L., and S. Dessai. 2010. "Robust Adaptation to Climate Change." *Weather* 65: 180–85.

Wilby, R. L., and I. Harris. 2006. "A Framework for Assessing Uncertainties in Climate Change Impacts: Low Flow Scenarios for the River Thames, UK." *Water Resources Research* 42 (2): W02419. doi:10.1029/2005WR004065.

Wilks, D. S. 1992. "Adapting Stochastic Weather Generation Algorithms for Climate Change Studies." *Climatic Change* 22 (1): 67–84.

———. 2002. "Realizations of Daily Weather in Forecast Seasonal Climate." *Journal of Hydrometeorology* 3 (2): 195–207.

———, and R. L. Wilby. 1999. "The Weather Generation Game: A Review of Stochastic Weather Models." *Progress in Physical Geography* 23 (3): 329–57.

Willems, P., and M. Vrac. 2011. "Statistical Precipitation Downscaling for Small-Scale Hydrological Impact Investigations of Climate Change." *Journal of Hydrology* 402 (3–4): 193–205.

Yates, D., S. Gangopadhyay, B. Rajagopalan, and K. Strzepek. 2003. "A Technique for Generating Regional Climate Scenarios Using a Nearest-Neighbor Algorithm." *Water Resources Research* 39 (7): 1199.

**CHAPTER 3**

# The Decision Tree
# Framework

## Introduction

The decision tree presented in this chapter is based on the basic principles of bottom-up climate risk assessment methodologies, such as developed in decision scaling and robust decision making (RDM). In preparing this approach, other existing decision support tools were reviewed, including those produced by the United Kingdom Climate Impacts Programme, the World Bank Climate Change Group, the United States Agency for International Development, and the World Resources Institute. Insights from this review were incorporated into this tool's design. An overview of the process is presented in figure 3.1.

This process is hierarchical, meaning that in each phase of the analysis, either the process ends because the climate risks have been adequately addressed or the process proceeds to the next phase to address remaining concerns. Phases 1 through 3 are elements of risk *assessment*. Phase 4 shifts to risk *management*. This chapter provides a detailed description of each phase in the decision scaling methodology.

The overarching objective of the decision tree framework is to provide a consistent, credible, and repeatable process for project managers to use to assess climate risks. It is designed such that effort expended is proportional to

**FIGURE 3.1 General Steps in the Decision Tree for Water Resources Projects**

**Phase 1: Project Screening**

- Stakeholder-defined performance indicators and risk thresholds are established.
- The proposed project is classified according to its climate sensitivity.
- Context analysis is performed using the Four C's framework, and potential climate vulnerabilities are described relative to potential vulnerabilities of other types.

**Phase 2: Initial Analysis**

- A water resources system model (at least a low-level approximation of the system) is developed, if not already available.
- An estimate of the magnitude of potential climate stressors relative to stressors of other types is developed through a rapid project scoping exercise and reported in a Climate Risk Statement.

**Phase 3: Climate Stress Test**

- A complete coupled hydrologic-economic water system model is built, if not already available.
- An exhaustive climate stress test is conducted to identify the climate sensitivity of the system, and is presented on a climate response map.
- The addition of historical data, paleoclimatology data, and general circulation model projections to the climate response map illustrates the risk to the system within the plausible range of climate changes relevant in the economic lifetime of the project.
- Projects shown to be insensitive to plausible climate changes are categorized as climate invulnerable in a Climate Risk Report.
- Climate response maps revealing ambiguity in the projected climate-related impacts are subjected to a credibility assessment.
- The robustness of the project to credible, plausible risks is examined.

**Phase 4: Climate Risk Management**

- If the robustness is unsatisfactory, and cannot be improved, alternatives are pursued.
- If the robustness can be easily and directly improved, adjustments are made to the project design, and the revised project is returned to the climate stress test.
- If doubt remains about the robustness of the project, and robustness improvements cannot be made with simple adjustments to design parameters, then advanced tools for decision making under uncertainty are employed using the ex post scenarios identified during Phase 3 and elaborated on in Phase 4.
- The results are presented in a Climate Risk Management Plan.

the climate sensitivity of the project in question. It is also designed to be implemented largely internally, although at some points outside expertise may be warranted. When outside expertise is needed, the framework provides clear guidance on the process that should be specified in the contract. Finally, the framework is expected to allow a project manager to feel confident that climate change risks have been assessed and addressed, and for management and the larger stakeholder community to agree.

Figure 3.2 presents a schematic workflow of the tool designed for water resources projects. The procedure consists of four hierarchical phases— Phase 1: Project Screening; Phase 2: Initial Analysis; Phase 3: Climate Stress Test; and Phase 4: Climate Risk Management. The overarching goal of the decision tree framework is to enable the project manager to confidently estimate the performance of the considered project, and to anticipate its impacts (positive and negative) on the natural and anthropocentric environment into which it would be introduced. More generic water resources system planning tools developed to accomplish this goal take climate (and other) information as input. The particular strength of the framework presented in this book is its handling of climate-related uncertainty. For situations in which climate information, among all sources of uncertain information, is inadequate for the task of evaluating project performance, the decision tree provides bottom-up procedures for clarifying the climate information.

Phase 1 consists of a well-defined, self-guided desktop screening of the project, to be conducted with the assistance of directed questions included on a Climate Screening Worksheet. The project manager would execute Phase 1 with little need for expert consultation. Projects that do not have significant climate sensitivities are excused from further analysis, and projects that do have potentially significant climate sensitivities may still be excused if those sensitivities are shown to be minor relative to other, larger, project-performance sensitivities. The judgment is made by the project planner and facilitated by guiding questions.

Projects classified as climate sensitive in Phase 1 move to Phase 2, which consists of an initial analysis, during which it is necessary to build a first-order approximate model of the proposed water system (if such a model does not already exist) and incorporate the expert input of stakeholders, regional managers, and others. Though not a thorough climate change risk assessment, the rapid scoping exercise is able to estimate a project's relative sensitivity to climate changes in the range projected by available general circulation models (GCMs) and indicate whether a more in-depth climate change risk assessment is required.

For projects with climate sensitivities that are significant relative to other, nonclimate, sources of performance sensitivities, a Phase 3 stress test is recommended. This step requires the use of a coupled hydrologic-economic analytical model to assess the climate sensitivity of the project in quantitative terms. Phase 3 would likely be performed by a qualified team of internal staff or expert consultants with knowledge of decision scaling applications.

Phases 1, 2, and 3 include points for exit from the decision tree, to be used when the climate information is deemed adequate (relative to other relevant sources of information) for the purpose of project evaluation. If significant and

## FIGURE 3.2 Decision Tree Schematic

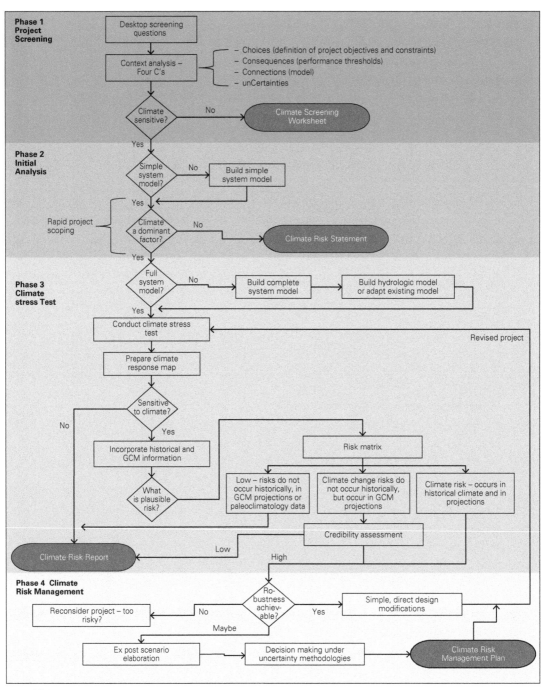

*Note:* GCM = general circulation model.

credible climate risk is identified, Phase 4 risk management is required. In this case, climate vulnerabilities are reduced through design modifications (or, in the extreme, the abandonment of the original design) using tools for decision making under uncertainty. Methods for dealing with decision-making uncertainty and, in some cases, further geophysical analysis, are needed. Chapter 5 discusses decision making under uncertainty in more depth.

For each phase, a product has been specified that would be appropriate for inclusion in the project folder and for presentation to reviewers to demonstrate that climate risks have been assessed according to an approved procedure. For example, Phase 1 results in a completed Climate Screening Worksheet, demonstrating the climate sensitivity (or lack thereof) of a given project. Phase 4, when necessary, results in an in-depth report, the Climate Risk Management Plan, that outlines the climate risks and proposed means of addressing them. Each step is defined in greater detail in the rest of this chapter.

## Phase 1: Project Screening

In Phase 1, a project is quickly screened to establish whether it is sensitive to climate factors.

### Entry into Phase 1

Phase 1 is the default starting point: all projects enter Phase 1.

### Description of Phase 1

The objective of project screening is to quickly assess and "clear" water resources projects that do not have climate sensitivities. To aid in this process, a Climate Screening Worksheet has been developed to guide the project evaluator through a series of screening questions. The proposed project is classified into one of only two categories: significantly climate sensitive or not significantly climate sensitive. The classification would be based on results from climate risk screening tools[1] complemented by the answers to the questions in the worksheet. Examples of complementary screening questions include the following (see the Climate Screening Worksheet in appendix B for a broader list of questions):

- What are the stakeholder-defined performance indicators (PIs) and risk thresholds?
- Is this a water infrastructure project?

- What is the proposed economic lifetime of the project?
- What discount rate is preferred (for example, a social-welfare or finance-equivalent rate; see box 3.1)?
- Is this a water policy adjustment, training session, or environmental or water resources study (without infrastructure)?
- Is this a hydrometeorological service project?
- Does the project use historical streamflow or climate data in the design?

The worksheet is designed to lead to a clear classification. In general, projects involving infrastructure would proceed to Phase 2, while more-policy-oriented projects, such as water sector reform, training activities, and

---

BOX 3.1

## Discount Rates

The discount rate is a particularly influential parameter in the ranking of investment alternatives. Higher discount rates result in lower economic design lives and deemphasize benefits accrued long into the future (for example, by future generations). Stern (2007) uses a consumption discount rate of 1.4 percent, while others (for example, Mendelsohn 2008; Nordhaus 2007) recommend higher discount rates—nearer 5 percent. (Nordhaus [2007], for instance, endorses a consumption discount rate of 4.3 percent.) The low consumption discount rate chosen by Stern results in a recommendation for a more aggressive policy to combat climate change, whereas the higher discount rate endorsed by Nordhaus results in a considerably more modest proposed effort.

In practice, the social discount rates used to evaluate the net benefits of proposed projects have varied widely, with developed nations typically applying a lower rate (3–7 percent) than developing nations (8–15 percent) (Zhuang et al. 2007). Organizations such as the World Bank, the Inter-American Development Bank, and the Asian Development Bank use a discount rate of 12 percent, although in some cases (such as

water supply projects), 10 percent is the norm. The essential rationale for these elevated discount rates stems from the high value of scarce capital in developing countries: projects consuming large amounts of capital are required to account for the opportunity cost of these financial resources, pushing up the expected rate of return.

Goulder and Williams (2012) distinguish between a *social-welfare-equivalent* discount rate, appropriate for determining whether a given policy would augment social welfare, and a *finance-equivalent* discount rate, suitable for determining whether the policy would offer a potential Pareto improvement.[a] The distinction underscores the need for active stakeholder participation in local determination of the discount rate. The relevant question to stakeholders is, for whom are the benefits of development projects intended (distributed in space and in time)? Though the deliberations through which social discount rates are chosen are disconnected from evaluations of the net present value of individual proposed projects, the conclusions reached go a long way toward establishing the attractiveness of the project under consideration.

a. A Pareto improvement helps at least one party to a negotiation, and harms none.

studies, would be classified as climate insensitive and reach the endpoint of the climate risk analysis; this latter category may, however, continue through the decision tree to analyze other sources of uncertainty.

### The Four C's

A key aspect of the Climate Screening Worksheet is the context analysis, which defines the bounds of the project being evaluated. For the purpose of this work, the context analysis is guided by a framework described as the Four C's: Choices, Consequences, Connections, and unCertainties. The project analyst, in consultation with stakeholders, will identify available project design choices; the consequences or performance metrics that will be used to evaluate the project's success; the connections through which choices and consequences are linked; and finally, the uncertainties that affect the conclusions that can be drawn about the profitability of the project. Through this analytical approach to assessing the project, the relative effects of the various uncertainties are expected to emerge.

**Product:** The **Climate Screening Worksheet** is a standardized document with a series of questions that result in the categorization of the project as either climate sensitive (leading to Phase 2) or not climate sensitive (leading to the end of the climate assessment process, and exit from Phase 1); it is approximately two pages long (see appendix B). The Climate Screening Worksheet is completed regardless of project categorization (climate sensitive or not). It is either submitted alone (as evidence of the climate insensitivity of the project), or included as part of more thorough climate assessment reports as described in relation to later phases of the decision tree.

### Exit from Decision Tree after Phase 1

If the Climate Screening Worksheet suggests that the proposed project has no measurable climate sensitivities, the project may exit the decision tree (see figure 3.3). In this case, a climate risk assessment is not necessary and the project manager can proceed with standard procedures for evaluating the project.[2]

## Phase 2: Initial Analysis

In Phase 2, a cursory analysis is conducted to determine whether the identified sensitivities to climate should be considered relatively insignificant compared with sensitivities to nonclimate factors.

FIGURE 3.3 **Phase 1 Entry and Exit Conditions**

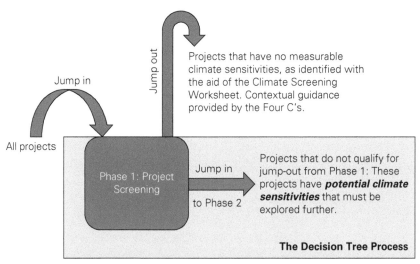

Note: Four C's are Choices, Consequences, Connections, and unCertainties. See text for discussion.

## Entry into Phase 2 from Phase 1

If the project is shown to have significant potential climate sensitivities in the Climate Screening Worksheet, the decision tree process proceeds from Phase 1 to Phase 2.

## Description of Phase 2

The intent of Phase 2 is to further explore the context of the project, and to identify cases in which climate sensitivities are present but deemed unlikely to be important relative to sensitivities of other types. Expert-led climate sensitivity analysis using either an existing water system model (typically developed as part of a prefeasibility study) or a simple water balance informed by available GCMs will serve to excuse many water resources projects from further in-depth climate risk assessment in Phases 3 and 4 of the decision tree.

A number of techniques are available for diagnosing the sensitivity of the system to climate changes relative to changes in other factors, ranging from simple one-factor-at-a-time sensitivity analysis (see the section on "Extensions to Traditional Decision Analysis" in chapter 5) to rapid project scoping (see the "Rapid Project Scoping" section in this chapter) to the "patient" rule induction method (PRIM) (Friedman and Fisher 1999).

Confronting Climate Uncertainty in Water Resources Planning and Project Design

PRIM can be used to identify the combinations of input parameters under which the system performs poorly. By systematically exploring all possible combinations of input parameters simultaneously, incremental changes in hydrologic system inputs can be ranked in significance relative to incremental changes in nonhydrologic inputs. If a PRIM-based analysis were to identify relatively significant system sensitivities to hydrologic changes within a range that might reasonably be expected to occur during the project's design life (for example, within 1 standard deviation of historical mean annual streamflow), those sensitivities would be explored in greater depth in a Phase 3 Climate Stress Test. In climate change risk assessment, PRIM is best applied as a screening tool to identify general system sensitivities, with more targeted explorations of the effects of potential changes in specific climate characteristics (mean temperature and precipitation, precipitation variability, seasonal shift, monsoon intensity and duration, timing and frequency of arrival of atmospheric rivers, and so forth) undertaken in Phase 3. Because PRIM is computationally intensive and involves a substantial measure of model development or customization, this section describes a less sophisticated and more direct method for characterizing the relative significance of various system sensitivities, known as rapid project scoping.

Much of the analysis will often have already been completed by external consultants as part of the prefeasibility study, and would thus not need to be repeated by the project manager at this stage. An important responsibility of the project manager is therefore to thoroughly review existing hydrologic studies completed at earlier phases of the evaluation of the proposed water resources system investment.

## Rapid Project Scoping

Rapid project scoping is based on a simplified evaluation of the hydrology and climate change projections (Grijsen 2014b), as described, for example, in figure 3.4. The scoping analysis is designed to be executable by technical staff using spreadsheet software (Excel, for instance), though in many cases this analysis would be more efficiently performed by external expert consultants. The analysis uses regressions to relate changes in climate parameters to changes in available water, and separate relationships to relate changes in available water to changes in system performance. Though the procedure outlined in figure 3.4 is not a thorough climate change risk assessment, it enables estimation of a project's relative sensitivity to climate changes in the range projected by available GCMs. If the potential risks to system performance are shown to be significant, further analysis may be

FIGURE 3.4 **Project Scoping Workflow for Phase 2**

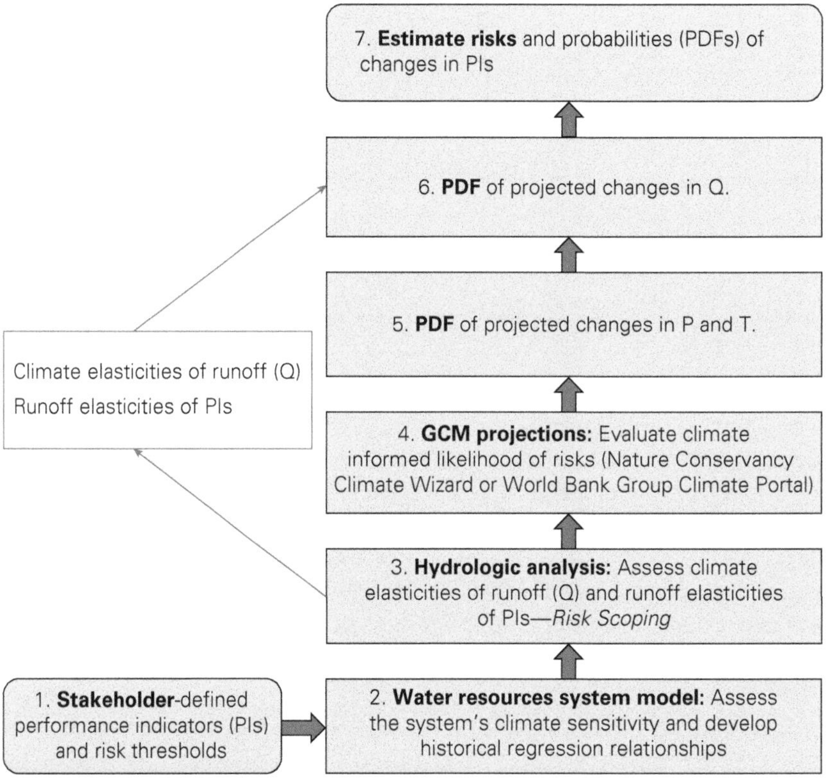

7. **Estimate risks** and probabilities (PDFs) of changes in PIs

6. **PDF** of projected changes in Q.

5. **PDF** of projected changes in P and T.

Climate elasticities of runoff (Q)
Runoff elasticities of PIs

4. **GCM projections:** Evaluate climate informed likelihood of risks (Nature Conservancy Climate Wizard or World Bank Group Climate Portal)

3. **Hydrologic analysis:** Assess climate elasticities of runoff (Q) and runoff elasticities of PIs—*Risk Scoping*

1. **Stakeholder**-defined performance indicators (PIs) and risk thresholds

2. **Water resources system model:** Assess the system's climate sensitivity and develop historical regression relationships

*Source:* Adapted from Grijsen (2014a).

*Note:* GCM = general circulation model; PDF = probability density function; P = precipitation; PI = performance indicator; Q = runoff; T = temperature.

justified in Phase 3 of the decision tree. If not, the project may be allowed to exit the decision tree.

The bulk of the effort in this phase is spent on water resources system modeling and analysis, to determine the runoff elasticity of selected PIs. If a water resources system model has not been developed, one will be needed for this analysis, even if it is only a basic description of essential system components using spreadsheet software. At this stage, publicly available hydrologic data will also need to be collected to create a baseline runoff record, and climate change projections will need to be acquired. A hydrologic model is required at this stage only if data from the historical record are of insufficient length or quality.

Effort on the order of a month or two of analysis for most projects will suffice, including all model development. If all required models are preexisting and available, the analysis may be accomplished much more quickly (in the range of two to three weeks). Using the system model, a first-cut exploration of the local historical hydrology and climate change projections translates into estimates of the likelihood that selected PIs will perform inadequately. It is useful to express all results in terms of relative (that is, percentage) changes. Relating percentage changes in PIs to percentage changes in runoff (as opposed to using absolute values) substantially reduces systemic errors in modeling, and redirects discussions with stakeholders away from disputable absolute numbers.

The rapid scoping exercise involves the following steps:

1. *Establish stakeholder-defined PIs* and acceptable risk levels (thresholds).
2. Develop (or acquire) a water resources model of the proposed system (or proposed modifications to existing system). Use the system model to *assess the response of the anthropocentric water system to varied hydrologic inputs*. This step is accomplished by *developing elasticities*, $\varepsilon_Q$, of basin PIs (GWh/year, firm power, irrigated area, minimum flow, navigation, and so forth) by parametrically varying basin runoff. These elasticities are applied to risk assessment in step 7. In addition to regression relationships, first-order approximate (for example, mass balance) hydrologic modeling may be required if the record of historical streamflow observations is inadequate. Example elasticity plots are presented in figure 3.5. As shown in figure 3.5, the rapid scoping exercise need not be conducted on a single design at a time; it can be conducted on numerous design options simultaneously (or on a single prefeasibility option and a small number of alternatives).
3. *Assess the response of the hydrologic system to varied climate inputs* by parametrically varying flow and climatic boundary conditions. This step is accomplished by first generating regression relationships between climate parameters and streamflow using time series of historical climate and streamflow data. In addition to regression relationships, first-order hydrologic modeling may be required. The primary purpose of this step is to generate a second set of elasticities—climate elasticities to changes in precipitation and temperature—based on (1) historical runoff (Q) data and (2) hydrometeorological or gridded data sets of precipitation (P) and temperature (T). These elasticities quantify the response of the hydrologic system (as represented by runoff and streamflow) to climate drivers within the bounds of the historical record.
4. *Use climate information* from several sources of climate projections (for example, from the Nature Conservancy's Climate Wizard or the Bank's Climate Change Knowledge Portal) to assess the likelihood of climate risks for

FIGURE 3.5 **Example of Elasticities of Basin Performance Metrics**

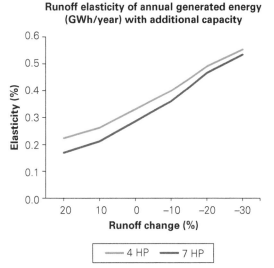

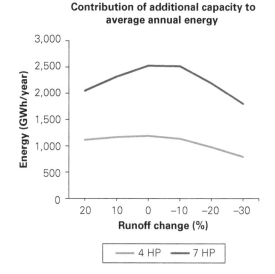

*Source:* Grijsen 2014a.

*Note:* GWh = gigawatt hour; HP = hydropower plants. These data apply to alternative designs using four and seven hydropower plants on the Sanaga River in Cameroon.

the specified investment horizons. GCM data should be complemented with analysis of glacial melt water and sea level rise risks, where appropriate.

5. *Develop probability density functions* (PDFs)[3] of projected changes in precipitation and temperature (see figure 3.6 for an example).
6. *Develop PDFs* of projected changes in streamflow (or available water) by way of the hydrologic analysis in step 3 (see figure 3.6 for illustration).
7. Using the information from step 6 in combination with the water resources system model (and associated elasticities) developed in step 2, *estimate the risks and probabilities* of changes in the PIs of concern. This estimation is made by developing PDFs of "projected" changes in PIs: $E[\Delta PI/PI_0] = \varepsilon_Q E[\Delta Q/Q_0]$, where $\varepsilon_Q$ may be a nonlinear function of $\Delta Q/Q_0$ (see figure 3.7 for an example).

An example of conclusion to the rapid project scoping, which results in exit from Phase 2, is presented for the Sanaga Basin, Cameroon, in Grijsen (2014a). In this case, rapid project scoping indicated that the total average energy generation of four run-of-the-river hydropower plants in the basin had an elasticity to flow (Q-elasticity) of 0.3 to 0.5, and that it was highly unlikely that the threshold for the stakeholder-defined PI (in this case a change in energy production relative to the baseline of > 20 percent) would be crossed by 2050 or 2080. There was a high probability that the economic

Confronting Climate Uncertainty in Water Resources Planning and Project Design

**FIGURE 3.6 Example of Changes in Precipitation, Temperature, and Runoff According to General Circulation Model Projections**

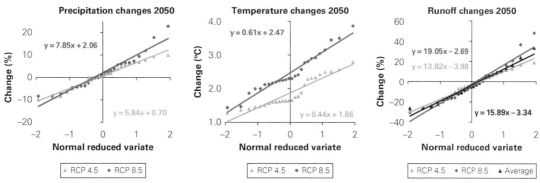

*Note:* GCM = general circulation model; RCP = representative concentration pathway. This illustration is for the Upper Niger Basin at Bamako, Mali. The symbol ▲ represents, respectively, precipitation, temperature, and runoff changes according to GCM climate projections for RCP 4.5 of the Coupled Model Intercomparison Project phase 5 (CMIP5) generation; the symbol ◆ represents those respective changes for RCP 8.5 of the CMIP5 generation. The reduced variate of the normal distribution is defined as $Z = (x - \mu)/\sigma$. $Z$ has a mean of zero and a standard deviation of 1, making it easier to visualize the effect of changes in the independent variables (precipitation, temperature, runoff) on the dependent variable (runoff, water system performance).

**FIGURE 3.7 Example of Changes to Selected Performance Indicators Associated with General Circulation Model Projections**

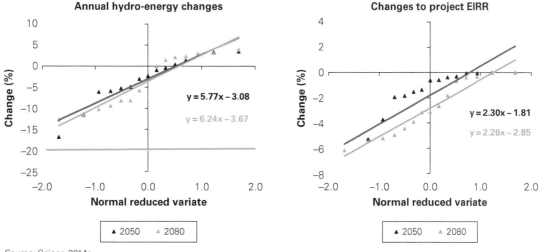

*Source:* Grijsen 2014a.

*Note:* EIRR = economic internal rate of return; GCM = general circulation model; PI = performance indicator. These data apply to the scenario of four hydropower plants on the Sanaga River in Cameroon. The symbol ▲ represents Coupled Model Intercomparison Project phase 3 (CMIP3) GCM climate projections for 2050; the symbol ▲ represents CMIP3 GCM climate projections for 2080.

internal rate of return (EIRR) of the Lom Pangar and Nachtigal dam projects on the Sanaga basin, specifically, would not change significantly as a result of climate change (< 5 percent of the projected 14.5 percent EIRR). In the worst case, the EIRR was estimated to fall to 13 percent from 14.5 percent, still within the range of attractive investments for the project sponsor. It was therefore concluded that the evaluated hydropower projects are economically robust and climate insensitive.

Climate risk screening and adaptation tools, such as the World Bank's Climate Change Knowledge Portal (http://sdwebx.worldbank.org/climateportal/index.cfm) and the Nature Conservancy's Climate Wizard (http://www.climatewizard.org/), are helpful tools for quickly identifying anticipated changes in temperature, precipitation, and other climate conditions relevant to the location of the planned water project. The United Nations Development Programme's (UNDP) Adaptation Learning Mechanism (http://www.adaptationlearning.net/) offers geographically targeted resources for climate change adaptation. If the proposed project is believed to be potentially sensitive to large decreases in precipitation, for example, these resources for climate information can inform the project manager of the likely magnitude of precipitation changes in the region of interest, allowing climate risks to be considered relative to risks of other types.

Additionally, as pointed out in step 2, the rapid scoping exercise need not be conducted on a single design at a time; the exercise can be conducted on numerous design options simultaneously (or on a single prefeasibility option and a small number of alternatives).

Completion of this analysis may require consultation with internal experts, or a regional manager or colleague. The expectation is that by completing the Four C's and the rapid scoping procedure presented in figure 3.4, the project manager will gain the clarity necessary to judge whether further analysis of the project for climate risks is warranted. If the analyst concludes that the project does not require further analysis, a Climate Risk Statement is completed (a template is provided for this statement in appendix B).

If sensitivity to climate is determined to be a significant factor in the project's expected performance, a more detailed model of the system's response to climate changes may be required. If the project cannot be excused from an in-depth climate stress test, the now intermediate Climate Risk Statement can be skipped, given that its content will be covered in the Phase 3 Climate Risk Report.

**Exit from Decision Tree after Phase 2**

If Phase 2 of the decision tree shows that the project has climate sensitivities, but that those sensitivities are small relative to sensitivities to uncertain factors of other types (for example, demographic or political factors; see figure 3.8), standard internal procedures for evaluating the project should be followed after completion of the Climate Risk Statement. For expensive or complex projects, it is recommended, given the potential sensitivities identified in Phase 1, that the project manager employ some extension of traditional decision analysis (example methods are summarized in "Extensions to Traditional Decision Analysis" in chapter 5) to assist in the selection of the best project design.

As was the case with exit from Phase 1, if potential nonclimate vulnerabilities have been shown in this phase to be much more significant than potential climate vulnerabilities, a nonclimate version of the stress test might be used to explore those vulnerabilities quantitatively. Such methods are not described in detail here.

**Product:** The **Climate Risk Statement** outlines the effects of uncertainty on the project and the expected relative effect of climate uncertainty in comparison with other uncertainties; it is normally fewer than five pages long. The statement should justify why further climate analysis

FIGURE 3.8 **Phase 2 Entry and Exit Conditions**

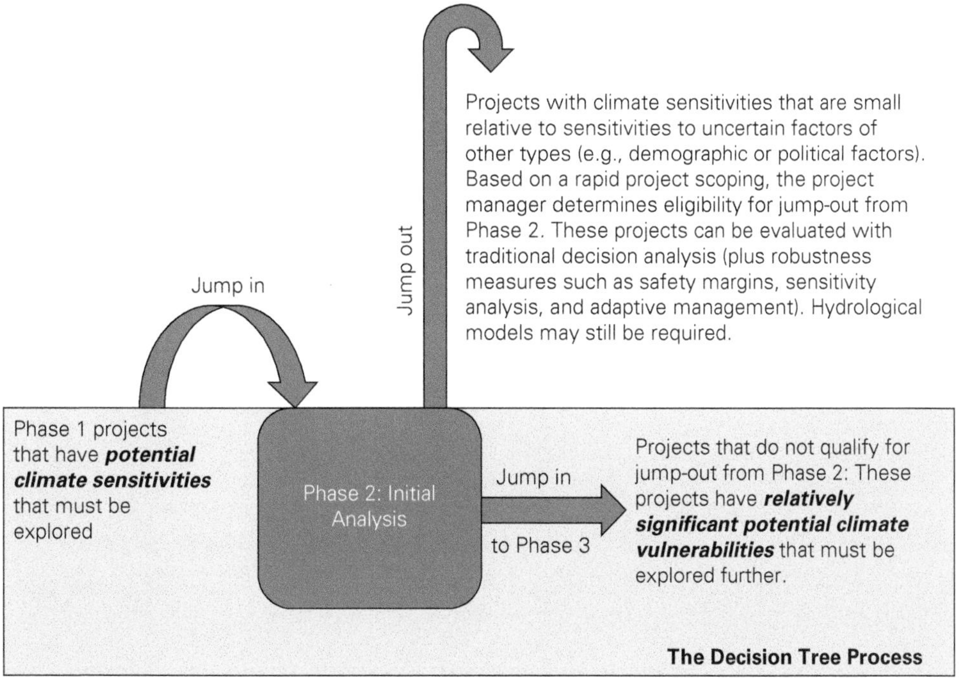

Projects with climate sensitivities that are small relative to sensitivities to uncertain factors of other types (e.g., demographic or political factors). Based on a rapid project scoping, the project manager determines eligibility for jump-out from Phase 2. These projects can be evaluated with traditional decision analysis (plus robustness measures such as safety margins, sensitivity analysis, and adaptive management). Hydrological models may still be required.

Jump in

Jump out

Phase 1 projects that have **potential climate sensitivities** that must be explored

Phase 2: Initial Analysis

Jump in to Phase 3

Projects that do not qualify for jump-out from Phase 2: These projects have **relatively significant potential climate vulnerabilities** that must be explored further.

**The Decision Tree Process**

is not required, for instance, because of the dominant effect of other uncertainties. A template is provided for the completion of this statement in appendix B.

The Climate Risk Statement is completed only if Phase 2 is exited; otherwise, the climate risks will be described as part of a more in-depth report on climate vulnerabilities (the Climate Risk Report of Phase 3 or the Climate Risk Management Plan of Phase 4).

## Phase 3: Climate Stress Test

In Phase 3, a project is subjected to a so-called climate stress test, as outlined in box 3.2.

### Enter Phase 3 from Phase 2

The Four C's analysis of Phase 1 will help identify the pathways through which the various uncertainties (climate and nonclimate) might affect the

---

**BOX 3.2**
### General Procedure for a Climate Stress Test

First, a weather generator is developed for the region of interest to produce numerous stochastic time series that preserve the variability and seasonal and spatial correlations of the historical record. These time series can be generated either by resampling directly from the historical record or by generating new time series based on the statistical characteristics of the historical record.

Next, the parameters are systematically changed to produce new sequences of weather variables (for example, precipitation) that exhibit a wide range of change in their characteristics (average amount, frequency, intensity, duration, and so forth). For instance, linear trends could be added to the precipitation and temperature of the numerous stochastic time series to simulate climate change on a range informed by the available downscaled general circulation models. An example of a method for conducting the stress test is provided in Steinschneider and Brown (2013).

Using the stochastic time series, the hydrologic and water resources system model is then run repeatedly for the entire period for many future climates, for each of the water system plans under consideration.

Finally, the performance of each proposed plan is evaluated for a range of future climate states and the results are presented on a climate response map. Some aspects of system performance that may be evaluated are benefit-cost ratio, total net benefits, economic rate of return, and violations of performance thresholds.

project. The initial analysis in Phase 2 will indicate whether the climate sensitivities identified in Phase 1 are relatively significant. If potential climate vulnerabilities are not insignificant relative to nonclimate risks, a Phase 3 analysis is begun.

Generally, infrastructure projects with design lives longer than 10–20 years, especially projects in geographic regions with high inter- or intra-annual climate variability, will require more thorough project scoping in Phase 3. More carefully constructed hydrologic and water resources system models will need to be constructed, with particular attention given to capturing potential changes or shifts in climate other than percentage changes in annual average temperature and precipitation.

## Description of Phase 3

Proposed projects entering Phase 3 have potential climate-related vulnerabilities significant enough relative to other risks that they cannot be dismissed. Phase 3 is the point at which the climate concerns are so prominent that they warrant heavier computational analysis. If a formal model of the natural, engineered, or socioeconomic system is not available at this point, it must be created to relate climate conditions to the impacts on PIs identified in Phase 2. The third phase of the decision tree is the first that is highly technical and will require an external expert consultant or qualified internal experts. Data availability, as well as cost and time constraints for the analysis, must be duly considered. The overall approach to a climate stress test is described in box 3.2.

To explore the climate sensitivity of a project, climate response functions may be developed by systematically varying climate conditions and recording changes in performance metrics. The risks to the system are exhaustively explored and identified by testing a wide variety of possible climate futures, beyond the narrow range available in a typical GCM analysis (Stainforth et al. 2007). A climate response map is a useful way to portray the sensitivities of the system to climate changes, as shown in figure 3.9. Figure 3.9 is taken from an evaluation of a run-of-the-river hydropower facility, which will be referred to throughout this book and explained in greater depth in chapter 4. For now, let it suffice to illustrate the main features of a standard climate response map.

A climate response map demonstrates the performance of a system across a wide range of possible climate states. In this example, the performance of a hydropower facility with a specific design capacity is being tested across the range of climates shown. Note that the climate space over which the system is tested includes and substantially exceeds both the historical climate

FIGURE 3.9 **Example of a Climate Response Map for a Proposed Run-of-the-River Hydropower Project**

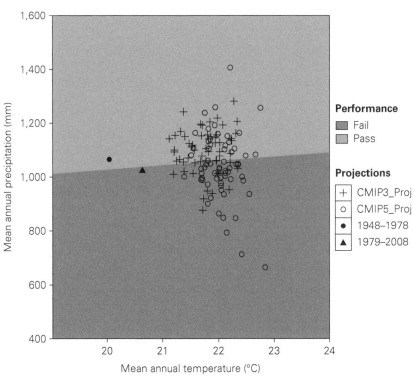

*Note:* CMIP = Coupled Model Intercomparison Project. The symbols ● and ▲ refer to annual averages for temperature and precipitation for the periods shown. Threshold = levelized cost regret of $500/GWh. Downscaled general circulation model values are 20-year averages from 2030 to 2050. The region of acceptable system performance is shown in green; the region of failure in red.

(demonstrating the recent trend of increasing temperature and decreasing precipitation) and all available downscaled GCMs (both phases 3 and 5 of the Coupled Model Intercomparison Project [CMIP3 and CMIP5]). A performance threshold of a levelized cost regret[4] of $500/GWh has been set. The climate response map identifies climate states that result in unacceptable performance relative to the threshold. In the red area (low precipitation and relatively high temperature), the levelized cost regret of hydropower generated by the prefeasibility design is higher than $500/GWh,[5] which would constitute a "failure." The system would perform "acceptably" in the green region. Examples of ex post scenarios that would require further attention, therefore, would be the red region and the part of the green region close to the failure threshold.

The climate stress test is conducted using a model of the system, typically driven by hydrologic time series (although other model drivers are possible). During the climate stress test, the water system model is exposed to a very large number of stochastically generated climate states. The tested climate states cover a range of possible futures much wider than those suggested by historical data and GCMs. It is recommended that the organization establish a range of climate change for each geographical region for consistency in reviews of project assessments. Data can be obtained from the World Bank's Climate Change Knowledge Portal for this purpose. The climate response map (figure 3.9) demonstrates the sensitivity of the system to a wide range of climate states.

If no vulnerabilities are revealed on the climate response map, a Climate Risk Report will explain the process used and the fact that the project has been designated robust to climate change (and thus exits the decision tree). If the climate response map does show project vulnerabilities within the tested range, the system is determined to be vulnerable to changes in climate.

*If the system is shown to be vulnerable to changes in climate, further analysis is required to determine whether the problematic conditions are likely to arise.* This step frames the available climate information with regard to its effect on the project. Data can also be obtained from the Bank's Climate Change Knowledge Portal for this purpose. The evaluation of the risk associated with the climate response map can be presented, for example,[6] by a simple count of GCM projections to which the project is robust (see figure 3.10) or a risk matrix, as shown in table 3.1. According to figure 3.10, only 22 of the 121 GCM runs (about one-sixth of them) project

FIGURE 3.10 **Downscaled General Circulation Model Count for Climate Response Map Shown in Figure 3.9**

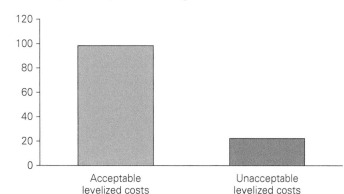

*Note:* The y-axis shows counts of general circulation model emission scenario projections.

**TABLE 3.1 Risk Matrix**

| | Likelihood | | |
|---|---|---|---|
| **Impact** | High impact and few GCM runs in vulnerable region of response surface; little or no historical indication that conditions are possible | High impact and many GCM runs in vulnerable region of response surface | High impact and many GCM runs in vulnerable region of response surface, and evidence from historical record and trajectory |
| | Medium impact and few GCM runs in vulnerable region of response surface; little or no historical indication that conditions are possible | Medium impact and many GCM runs in vulnerable region of response surface | Medium impact and many GCM runs in vulnerable region of response surface, and evidence from historical record and trajectory |
| | Low impact and few GCM runs in vulnerable region of response surface; little or no historical indication that conditions are possible | Low impact and many GCM runs in vulnerable region of response surface | Low impact and many GCM runs in vulnerable region of response surface, and evidence from historical record and trajectory |

*Note:* GCM = general circulation model. Dark green indicates relatively low risk; red indicates relatively high risk.

future climate states in which the levelized cost regret of this project exceeds the threshold. However, figure 3.10 does not offer any information about the magnitude of the failure, and though relative magnitudes can be inferred from figure 3.9, further analysis is required to quantitatively describe the magnitude of the risk.

Regardless of the magnitude, if the result of the risk matrix is that the impacts are unlikely to occur (for example, not a single projection indicates the conditions are likely and the conditions have not occurred historically or in the paleoclimatology record), the Climate Risk Report is completed (see appendix B). The report should explain the climate stress test to which the system model was subjected and conclude that detrimental impacts are unlikely to occur. Completion of the Climate Risk Report marks completion of the decision tree process—the project exits.

If the risks to project performance are very high (that is, both the historical record and some climate projections indicate that the conditions are possible), methods to improve the robustness of the project to those climate impacts must be evaluated. If, however, the impact to the project is in doubt (for example, the project is vulnerable to a climate state indicated by some GCM projections, but not to others), an assessment of the credibility of each data source should be conducted. For medium or high probability of impact (regardless of the magnitude of impact), the question of whether the robustness of the project can be improved must be asked.

The example presented in this section would have directed the project manager to Phase 4, Management of Climate Risks.[7] Not all projects entering Phase 3 necessarily move to Phase 4. In fact, many projects undergoing a Phase 3 Climate Stress Test will be shown not to have climate vulnerabilities that require adaptation strategies in Phase 4 Climate Risk Management; this is precisely one of the strengths of the Phase 3 Climate Stress Test. It is the responsibility of the project manager, the stakeholders, and the project evaluators to conduct the Phase 3 assessment in such a way that climate-related system vulnerabilities are described in the context of their *relative* significance, so that relatively small risks do not receive analytical attention out of proportion to their anticipated effect on the PIs of concern.

### Exit from Decision Tree after Phase 3

Projects exiting Phase 3 (and thus not entering Phase 4) will have been subjected to a climate stress test that showed the particular project design *not* to have substantive vulnerabilities in the climate range that might reasonably be expected to occur in the lifetime of the project (see figure 3.11). Therefore, advanced (and computationally expensive) tools for decision making under uncertainty are not required at this stage. However, it is recommended that any cost-benefit analyses done for these projects (which have been shown in Phase 2 to have significant climate sensitivities) give careful attention to the inclusion of safety margins and sensitivity analysis, as discussed in "Extensions to Traditional Decision Analysis" in chapter 5. Stochastic optimization procedures, such as benefit-cost analysis under uncertainty, multiobjective robust optimization, and real options analysis (discussed in "Stochastic Optimization" in chapter 5), would also be appropriate at this point.

**Product:** The **Climate Risk Report** details the climate stress test analysis process and its results. The sensitivity of the project is described and the

FIGURE 3.11 **Phase 3 Entry and Exit Conditions**

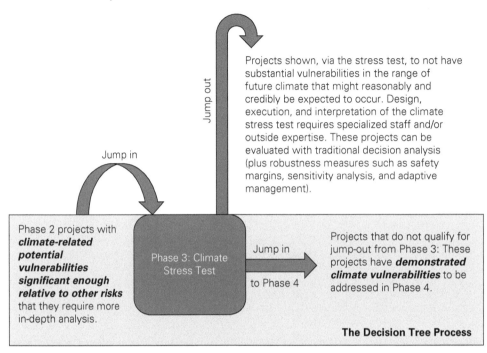

plausible risks associated with problematic conditions are explained, with the likelihood of each risk described based on the range of available climate information sources. A plan for addressing any climate risks that do emerge should be included. The suggested length of the report is 10–20 pages.

## Phase 4: Climate Risk Management

Phase 4 centers on the selection of tools to manage the climate risks identified in Phase 3 *under conditions of uncertainty*, once it has been established that the robustness of the project to a plausible range of climate change cannot be further improved. A project may also be considered too risky and be abandoned altogether at this stage.

### Entry into Phase 4 from Phase 3

Projects entering Phase 4 have substantial vulnerabilities in the range of climate that might reasonably be expected to occur in the lifetime of the project, as demonstrated in the Phase 3 climate stress test.

## Description of Phase 4

This step guides the project manager through selection of the tools available for managing the climate risks identified in Phase 3. The input to Phase 4 is a project that has been shown to have worrisome vulnerabilities to climate change that might reasonably be expected to occur within the lifetime of the project. The relevant question in Phase 4 is whether the robustness of the project to the plausible range of climate change can be improved. If the robustness of the project can be improved, the revised project should be subjected to the Phase 3 Climate Stress Test. If doubt remains about the ability to substantially improve the robustness of the project, methodologies for decision making under uncertainty should be used and reported on in the Climate Risk Management Plan. Some considerations regarding these methodologies are described in the "Risk Management Tools" section of chapter 5. In some cases, such as projects affected by uncertainty associated with changes in snowmelt or glacier melt, further geophysical analysis may also be warranted.

Given that the occurrence of a particular climate change scenario cannot be anticipated, preferred adaptation strategies should be robust across as wide a range of potential futures as possible. Hallegatte et al. (2012) provide a summary of four methodologies that have been used to support decisions under uncertainty: cost-benefit analysis under uncertainty (for example, Arrow et al. 1996), real options analysis (for example, Arrow and Fisher 1974; Henry 1974; Ranger et al. 2010), RDM (for example, Lempert and Schlesinger 2000; Lempert et al. 2006), and decision scaling, as discussed throughout this book. As described by Hallegatte et al. (2012), the latter two methods are both "context-first," robust decision approaches that differ mostly in the particular analytic tools they use and their relative emphasis on climate versus the combination of climate and socioeconomic uncertainties. The first two methods have routinely been incorporated into such robust analyses as a means of evaluating strategies.

The "Risk Management Tools" section of chapter 5 expands on those options, adding information gap decision theory (Ben-Haim 2006), dynamic adaptive policy pathways (Haasnoot et al. 2013), stochastic optimization (Loucks, Stedinger, and Haith 1981), and multiobjective robust optimization (Mulvey, Vanderbei, and Zenios 1995; Ray et al. 2014; Watkins and McKinney 1997) to the suggested techniques for decision making under uncertainty. Robust optimization and many-objective optimization (Reed and Minsker 2004) are sister techniques, and ideologically the same. In recent years, multiobjective decision-making approaches have been incorporated into RDM (Kasprzyk et al. 2013), thereby expanding beyond an optimization framework. Safety margins and sensitivity analysis, as described in "Extensions to Traditional Decision Analysis" in chapter 5, play essential roles in decision making under uncertainty, but are more directly applicable to the levels of climate sensitivity encountered in Phase 2.

Techniques that emphasize optimality are not recommended for decision making under uncertainty. Rather, preferred techniques aim at robustness to a wide range of futures (Brown and Wilby 2012; Kasprzyk et al. 2013; Lempert et al. 2006; Prudhomme et al. 2010; Ray et al. 2014; Wilby and Dessai 2010) or adaptive management techniques such as dynamic adaptive policy pathways or real options analysis, which add flexibility to incrementally adapt to a wide range of futures (Adger, Arnell, and Tomkins 2005; HMT and Defra 2009; Jeuland and Whittington 2014; Ranger et al. 2010; Rosenzweig and Solecki 2014).

Rosenhead (1989) describes robustness as a particular perspective on flexibility. As philosophical approaches to model development, robustness and flexibility and adaptability are founded on slightly different premises. Techniques that aim at *robustness* skew toward the conservative because they seek solutions that perform satisfactorily even in unknown future conditions significantly worse than the expected. Techniques that emphasize adaptability do not necessarily recommend water system configurations that perform satisfactorily in the worst case, but hold open the option to upgrade the system if, over time, it begins to look like the worst case is more likely. However, whereas robustness is a decision criterion that needs to be expressed with respect to some performance metric (as discussed in "The Concept of Robustness" in chapter 5), adaptability is better understood as an attribute of a strategy, and is not expressed with respect to any particular performance metric, but rather is a means to achieve robustness (or optimality). The best water systems plans therefore aim at robustness by way of flexible and adaptable increments. This approach is demonstrated in the recent trend toward analytic methods that seek robust adaptive strategies (Walker and Marchau 2003; Lempert and Groves 2010; Groves et al. 2013; Haasnoot et al. 2013).

## Jump Back to Phase 3 from Phase 4

Projects modified in Phase 4 to decrease vulnerabilities to climate change, either by way of simple, direct design modifications or analysis using advanced tools for decision making under uncertainty, are resubmitted to the Phase 3 Climate Stress Test.

## Exit from Decision Tree after Phase 4

If improvements cannot be made, the project may be deemed too risky (confidence in the acceptable level of net benefits of the project cannot be claimed), and be abandoned in favor of an alternative (the "do-nothing" alternative being a viable option at this point).

**Product: Climate Risk Management Plan**—a project that reaches Phase 4 will have considerable climate vulnerabilities that must be addressed. Each risk management plan will be unique to the project considered but would ideally incorporate both adaptability/flexibility and robustness components. The plan should detail the climate risks faced and the means to address those risks. In some cases, the risks may be judged by consulted experts to be acceptable, without taking additional steps. In other cases, modifications might be proposed to ensure that the occurrence of certain climate conditions would not cause the project to fail in its objectives. The length of the report may vary, but is likely to be more than 20 pages.

FIGURE 3.12 **Phase 4 Entry and Exit Conditions**

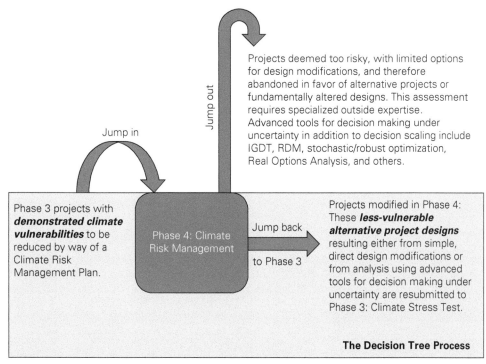

Jump out

Jump in

Projects deemed too risky, with limited options for design modifications, and therefore abandoned in favor of alternative projects or fundamentally altered designs. This assessment requires specialized outside expertise. Advanced tools for decision making under uncertainty in addition to decision scaling include IGDT, RDM, stochastic/robust optimization, Real Options Analysis, and others.

Phase 3 projects with **demonstrated climate vulnerabilities** to be reduced by way of a Climate Risk Management Plan.

Phase 4: Climate Risk Management

Jump back to Phase 3

Projects modified in Phase 4: These **less-vulnerable alternative project designs** resulting either from simple, direct design modifications or from analysis using advanced tools for decision making under uncertainty are resubmitted to Phase 3: Climate Stress Test.

**The Decision Tree Process**

Note: IGDT = information gap decision theory; RDM = robust decision making.

The Climate Risk Report is completed only in the event of exit from Phase 3. However, the goal of the decision tree is to achieve project designs with low vulnerabilities (high robustness) to climate change. Therefore, projects modified in Phase 4 (unless they are abandoned during Phase 4) are resubjected to the Phase 3 Climate Stress Test (see figure 3.12). If the Phase 4 design modifications were sufficient to successfully pass the project out of the decision tree through Phase 3, a Climate Risk Management Plan is included as part of the Climate Risk Report.

## Notes

1. For example, the World Bank's Climate Policy and Finance team has developed a set of climate risk screening tools.
2. Potentially, such procedures should include nonclimate versions of Phases 2 and 3 (and in some cases Phase 4), through which climate-insensitive projects are subjected to nonclimate stress tests, with risk assessment effort expended in proportion to the consequences of project failure. Nonclimate versions of the stress test are sensitivity analyses that involve straightforward modifications to the climate stress test described in this book, but in the interest of clarity are not presented here.
3. A PDF of a variable is a function that describes the relative likelihood that this variable will take on a given value.
4. Levelized cost is a form of cost-benefit analysis in which amortized annual cost is divided by annual average hydropower generated during the economic lifetime of the project. It represents the amount of money that must be charged per unit of energy to break even on the investment of building and operating the power plant. Levelized cost regret is the difference between the levelized cost for a particular design and the levelized cost for the lowest-cost design in a particular future. A robust project design is one that has low relative levelized cost regret across the considered range of climate futures. (See chapter 4 for more detail regarding the calculation and significance of this performance metric, as well as a discussion of the concept of "regret.")
5. The example is presented as an illustration of the decision tree procedure, and is not intended to be instructive with regard to the scale of benefits and costs at which real-world decisions are affected. For example, the cost regret threshold of $500/GWh ($0.0005/kWh) presented in figure 3.9 is unrealistically low; however, it is pedagogically useful. This chapter (and chapter 4) will carry forward the analysis at these scales, understanding that in reality the performance thresholds would be set at much higher values of cost regret.
6. GCM count is only one method for assigning probabilities to future climate states and may not be the best method available. It is shown here for its directness and simplicity, but falsely assumes the independence of each GCM result without acknowledging the common ancestry (clustering) of many GCM

assumptions. For this example, only 23 separate GCMs were run, using a number of different CMIP3 (A1B, A2, B1) and CMIP5 (RCP 4.5, RCP 8.5) emission scenarios. As will be described in chapter 4, a more scientifically defensible approach would be to assign probabilities to future climate states based on the fraction of the total projected future climate domains encapsulated by the particular climate state under consideration, using "climate change envelopes" (Stainforth et al. 2007).

7. With the reiterated caveat that, in reality, the magnitude of the risk threshold set for the example, $500/GWh, is very low, and would likely not be of concern to the hydropower project manager planning this project.

## References

Adger, N. A., N. A. Arnell, and E. L. Tomkins. 2005. "Successful Adaptation to Climate Change across Scales." *Global Environmental Change* 15: 77–86.

Arrow, K. J., M. Cropper, G. C. Eads, R. W Hahn, L. B. Lave, R. G. Noll, P. R. Portnoy, and others. 1996. *Benefit-Cost Analysis in Environmental, Health, and Safety Regulation*. Washington, DC: AEI Press.

Arrow, K. J., and A. Fisher. 1974. "Environmental Preservation, Uncertainty, and Irreversibility." *Quarterly Journal of Economics* 88 (2): 312–19.

Ben-Haim, Y. 2006. *Info-Gap Decision Theory: Decisions under Severe Uncertainty*, 2nd ed. London: Academic Press.

Brown, C., and R. L. Wilby. 2012. "An Alternate Approach to Assessing Climate Risks." *EOS, Transactions, American Geophysical Union* 92 (41): 401–12.

Friedman, J., and N. Fisher. 1999. "Bump Hunting in High-Dimensional Data." *Statistics and Computing* 9 (2): 123–43.

Goulder, L. H., and R. C. Williams III. 2012. "The Choice of Discount Rate for Climate Change Policy Evaluation." *Climate Change Economics* 3 (4): 1–18.

Grijsen, J. 2014a. "Climate Informed Decision Support Tools for Sustainable Water Management." Stockholm World Water Week—AGWA/WB seminar, Stockholm, Sweden, September 4.

Grijsen, J. 2014b. "Understanding the Impact of Climate Change on Hydropower: The Case of Cameroon." Report 87913, Africa Energy Practice, World Bank, Washington, DC.

Groves, D. G., J. R. Fischbach, E. Bloom, D. Knopman, and R. Keefe. 2013. *Adapting to a Changing Colorado River: Making Future Water Deliveries More Reliable through Robust Management Strategies*. Santa Monica, CA: RAND Corporation.

Haasnoot, M., J. H. Kwakkel, W. E. Walker, and J. ter Maat. 2013. "Dynamic Adaptive Policy Pathways: A Method for Crafting Robust Decisions for a Deeply Uncertain World." *Global Environmental Change—Human and Policy Dimensions* 23 (2): 485–98.

Hallegatte, S., A. Shah, C. Lempert, C. Brown, and S. Gill. 2012. "Investment Decision Making under Deep Uncertainty: Application to Climate Change." Policy Research Working Paper 6193, World Bank, Washington, DC.

Henry, C. 1974. "Investment Decisions under Uncertainty: The Irreversibility Effect." *American Economic Review* 64 (6): 1006–12.

HMT and Defra (HM Treasury and the Department for Environment, Food and Rural Affairs). 2009. *Accounting for the Effects of Climate Change: Supplementary Green Book Guidance.* London

Kasprzyk, J. R., S. Nataraj, P. M. Reed, and R. J. Lempert. 2013. "Many Objective Robust Decision Making for Complex Environmental Systems Undergoing Change." *Environmental Modelling and Software* 42 (April): 55–71.

Jeuland, M., and D. Whittington. 2014. "Water Resources Planning under Climate Change: Assessing the Robustness of Real Options for the Blue Nile." *Water Resources Research* 50 (3): 2086–107.

Lempert, R. J., and D. G. Groves. 2010. "Identifying and Evaluating Robust Adaptive Policy Responses to Climate Change for Water Management Agencies in the American West." *Technological Forecasting and Social Change* 77 (6): 960–74.

Lempert, R. J., D. G. Groves, S. W. Popper, and S. C. Bankes. 2006. "A General, Analytic Method for Generating Robust Strategies and Narrative Scenarios." *Management Science* 52 (4): 514–28.

Lempert, R. J., and M. E. Schlesinger. 2000. "Robust Strategies for Abating Climate Change—An Editorial Essay." *Climatic Change* 45 (3–4): 387–401.

Loucks, D. P., J. R. Stedinger, and D. A. Haith. 1981. *Water Resource Systems Planning and Analysis.* Englewood Cliffs, New Jersey: Prentice Hall.

Mendelsohn, R. O. 2008. "Is the Stern Review an Economic Analysis?" *Review of Environmental Economics and Policy* 2 (1): 45–60.

Mulvey, J. M., R. J. Vanderbei, and S. A. Zenios. 1995. "Robust Optimization of Large-Scale Systems." *Operations Research* 43 (2): 264–81.

Nordhaus, W. 2007. "A Review of the Stern Review on the Economics of Climate Change." *Journal of Economic Literature* 45 (September): 686–702.

Prudhomme, C., R. L. Wilby, S. Crooks, A. L. Kay, and N. S. Reynard. 2010. "Scenario-Neutral Approach to Climate Change Impact Studies: Application to Flood Risk." *Journal of Hydrology* 390 (3–4): 198–209.

Ranger, N., A. Millner, S. Dietz, S. Fankhauser, A. Lopez, and G. Ruta. 2010. "Adaptation in the UK: A Decision-Making Process." Grantham Research Institute on Climate Change and the Environment and Center for Climate Change and Economic Policy.

Ray, P. A., D. W. Watkins Jr., R. M. Vogel, and P. H. Kirshen. 2014. "A Performance-Based Evaluation of an Improved Robust Optimization Formulation." *Journal of Water Resources Planning and Management* 140 (6). doi:10.1061/(ASCE)WR.1943-5452.0000389.

Reed, P. M., and B. S. Minsker. 2004. "Striking the Balance: Long-Term Groundwater Modeling Design for Conflicting Objectives." *Journal of Water Resources Planning and Management* 130 (2): 140–49.

Rosenzweig, C., and W. D. Solecki. 2014. "Hurricane Sandy and Adaptation Pathways in New York: Lessons from a First-Responder City." *Global Environmental Change* 28 (September): 395–408.

Stainforth, D. A., T. E. Downing, R. Washington, A. Lopez, and M. New. 2007. "Issues in the Interpretation of Climate Model Ensembles to Inform Decisions." *Philosophical Transactions of the Royal Society A: Mathematical Physical and Engineering Sciences* 365 (1857): 2163–77.

Steinschneider, S., and C. Brown. 2013. "A Semiparametric Multivariate, Multi-Site Weather Generator with Low-Frequency Variability for Use in Climate Risk Assessments." *Water Resources Research* 49 (11): 7205–20.

Stern, N. 2007. *The Economics of Climate Change: The Stern Review*. Cambridge, U.K., and New York: Cambridge University Press.

Walker, W., and V. Marchau. 2003. "Dealing with Uncertainty in Policy Analysis and Policy-Making." *Integrated Assessment* 4 (1): 1–4.

Watkins, D. W., and D. C. McKinney. 1997. "Finding Robust Solutions to Water Resources Problems." *Journal of Water Resources Planning and Management* 123 (1): 49–58.

Wilby, R. L., and S. Dessai. 2010. "Robust Adaptation to Climate Change." *Weather* 65 (7): 180–85.

Zhuang, J., Z. Liang, T. Lin, and F. De Guzman. 2007. "Theory and Practice in the Choice of Social Discount Rate for Cost-Benefit Analysis: A Survey." Economics and Research Department Working Paper 94, Asian Development Bank, Manila.

**CHAPTER 4**

# Example Application: Run-of-the-River Hydropower

## Introduction

This chapter presents an application of the decision tree to a hypothetical example, a run-of-the-river hydropower project in Sub-Saharan Africa.[1] The example is presented to illustrate the decision tree procedure, and is not intended to be instructive with regard to the scale of benefits and costs at which real-world decisions are affected. For example, the cost regret threshold of \$500/GWh (\$0.0005/kWh) presented in figure 3.9, introduced in chapter 3, is unrealistically low; however, it is pedagogically useful. This chapter will carry forward the analysis at these scales, understanding that in reality the performance thresholds would be set much higher.

   The example run-of-the-river hydropower project involves the diversion of flow from two rivers through two small, concrete intake dams, and two identically sized tunnels (North and South) to a single hydropower turbine complex. The capacity of the hydropower plant originally proposed was 90 megawatts (31 cubic meters per second [$m^3/s$] turbine capacity). It has since been suggested that the design capacity could be increased to 200 megawatts. As will be demonstrated, the performance of the proposed hydropower facility is sensitive to changes in annual precipitation, but not particularly sensitive to temperature changes. This case study explores the effect of changes in future climate, both precipitation and temperature, on the performance of a number of proposed hydropower systems of various capacities.

## Phase 1: Project Screening

The proposed project is an infrastructure project with design life longer than 20 years in a part of the world with high seasonal and interannual precipitation variability. Based on that information alone, the Climate Screening Worksheet explains that the project is likely to continue to Phase 2 of the decision tree. However, climate sensitivities should be explored in context using the Four C's for guidance.

### The Four C's

Following the Four C's framework, the **Choices** available are to build a hydropower facility with a capacity of 31 m³/s, to build a hydropower facility with an alternate (larger or smaller) capacity, or not to build a hydropower facility. To understand the **Consequences**, the performance metrics must first be defined. As will be explained in greater depth in subsequent sections, the project is evaluated according to its cost efficiency (specifically, the levelized cost) and its total net benefits. Undesirable consequences, therefore, are either a poor benefit-cost ratio or low total net benefits. Consequences are understood relative to thresholds. If, for a given future climate state, the levelized cost is projected to be higher than some threshold levelized cost (for example, the levelized cost of the next-best-performing project design), or the total net benefits are less than zero, the project design would be considered in a state of failure. The project is being evaluated in isolation, with no **Connections** to other sectors and no other water projects explicitly modeled. As for **unCertainties**, demographic and economic factors are expected to affect the performance of the proposed system. However, because this is a water infrastructure project with a long design life in a part of the world that has high climate variability, it is reasonable to assume that the primary uncertainty in this evaluation would be the effect of a changing climate. The Four C's framework therefore points to the need for a rapid project scoping as part of a Phase 2 Initial Analysis. As part of Phase 2, performance thresholds are established in the next section.

## Phase 2: Initial Analysis

Figure 4.1 shows downscaled climate change projections between 2030 and 2050 for the region in which the proposed hydropower project would be located. Historical average precipitation and temperature for the period from 1948 to 2008 are about 1,020 millimeters (mm) and 20.2 degrees Celsius (°C), respectively.

**FIGURE 4.1  Downscaled Climate Change Projections for Region of Proposed Hydropower Project**

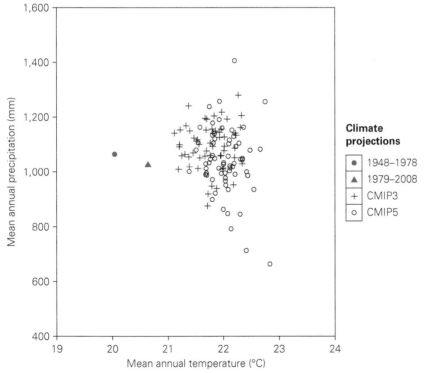

*Note:* CMIP = Coupled Model Intercomparison Project; mm = millimeters. The symbols ● and ▲ refer to annual averages (for temperature and precipitation) for the periods shown.

Figure 4.1 shows that between 1948 and 1978, the annual average temperature and precipitation were approximately 19.9°C and 1,040 mm, respectively. Between 1979 and 2008, the average annual temperature shifted up to 20.5°C, and the average annual precipitation shifted down to approximately 1,000 mm. The Coupled Model Intercomparison Project Phase 3 (CMIP3) general circulation model (GCM) results project an increase in temperature in this region of approximately 1.5°C (to 21.7°C) by 2040. The median CMIP5 GCM temperature projection is 22.0°C. Whereas CMIP3 precipitation projections range from approximately 850 to 1,250 mm/year, CMIP5 precipitation projections are in a range of about 650 to 1,325 mm/year, a range increase of almost 70 percent.

In a typical application of the decision tree, Phase 2 Initial Analysis is recommended to determine whether an in-depth Phase 3 Climate Risk Assessment is necessary; however, a Phase 2 project scoping exercise was not performed for this example.

### Performance Metrics

Two performance metrics were considered for this example: levelized cost and total net benefit. Elasticities of these two performance indicators to changes in streamflow, $\varepsilon_Q$, are used to estimate the risks to system performance of changes in climate. Using these performance metrics, three decision rules were evaluated in this example: minimum levelized cost regret, maximum probability of zero levelized cost, and maximum total net benefits.

*Levelized cost* is a cost-benefit ratio in which amortized annual cost is divided by annual average hydropower generated during the economic lifetime of the project. In this example, a 5 percent discount rate was used to amortize capital costs. See box 3.1 for a discussion of the effect of discount rate uncertainty on water system risk assessment. For each design capacity, annual operation and maintenance costs are a specified fraction of annualized capital costs, regardless of climate future (but specific to each design capacity).

$$Levelized\ cost = \frac{Amortized\ annual\ cost}{Annual\ average\ hydropower\ generated} \left[ \frac{\$}{kWh} \right] \quad (4.1)$$

*Levelized cost regret* is the difference between the levelized cost for a particular design and the levelized cost for the lowest-cost design in a particular future. A robust project design is one that has a relatively low levelized cost regret across the considered range of climate futures.

## Phase 3: Climate Stress Test

Typically, a quantitative Phase 2 Initial Analysis (for example, a rapid project scoping exercise) would be used to indicate the necessity of a Phase 3 Climate Stress Test. However, given that a quantitative initial analysis was not performed for this example, the more qualitative results of the Four C's assessment are used as guidance on the anticipated value added by a climate stress test. In this case, the Four C's framework suggests that a Phase 3 Climate Stress Test is probably warranted.

### Weather-Generator Procedure

The goals of the weather generator are to generate easily reproducible stochastic time series for the local system that preserve spatial and temporal correlations between sites (and also preserve temporal correlations between temperature and precipitation), and preserve low-frequency (interannual) variability. In this case, low-frequency variability was only explicitly preserved for the precipitation signal. The expectation is that the precipitation signal propagates to temperature, though low-frequency variability in temperature is not explicitly modeled. In context, low-frequency variability in temperature is unlikely to significantly affect system performance, and the added complexity of multivariate correlation was deemed to not be worth the potential marginal improvement in signal preservation.

A wavelet autoregressive model (WARM) was developed to identify low-frequency, interannual variability in the aggregated (average of two climate stations) annual time series of average monthly precipitation. Then, using the low-frequency statistics, many realizations (approximately 10,000) of 50-year time series of annual precipitation were generated, each of which was faithful to the low-frequency statistics of magnitude (mean, standard deviation, and skew) and frequency patterns (in this case, with a statistically significant signal at six years) in the historical record. Following a heuristic approach suggested by Lall and Sharma (1996), a $k$-Nearest Neighbor ($k$-NN) analysis was used for spatial and monthly disaggregation of stochastically generated annual time series. The $k$-NN procedure identifies a subset of years with annual precipitation totals nearest in value to the annual precipitation value of each particular stochastic realization. In this case, the subset consists of $k = 7$ historical years. One among the subset of seven historical years was next randomly selected, with highest likelihood of sampling the year with the annual precipitation value nearest the annual precipitation value of the stochastically resampled year. For the sampled year of the historical record, spatial disaggregation was achieved simply by substituting back into the new stochastic time series the actual climate values from the two climate stations. The two stations contain monthly values of mean precipitation, and minimum, maximum, and mean temperature, thus preserving spatial and temporal correlations.

This particular strategy for stochastic weather generation is but one among many possible for this application. It was chosen for its simplicity and because it perfectly preserves spatial and intra-annual correlations; however, the resampled precipitation variability cannot be greater than that observed in the historical record. In fact, it is almost guaranteed that the variability of the resampled subset of historical annual precipitation values will be lower than the historical, total variability. An alternative approach would be to approximately preserve spatial and intra-annual correlations by using the $k$-NN procedure to resample years with total precipitation similar to that of the WARM-generated annual precipitation values, but retaining the WARM-generated totals and disaggregating spatially and subannually by using the disaggregation factors relevant to each resampled year and site. Although this alternative approach would only approximately preserve spatial and intra-annual correlations, it would better represent the potentially increased range of future precipitation variability. The difference in methods is not expected to be particularly significant in this case.

### Climate Stress Test Procedure

When developing the climate response surface, it is advantageous to run the system model (in this case, the Water Evaluation and Planning [WEAP] model developed by the Stockholm Environment Institute) as many times as possible, both to trace out as wide a space of possible future climate as possible, and to fill each potential future climate space with as many stochastic runs as possible. However, trade-offs must be considered between comprehensiveness and computational burden. For this analysis, the WEAP model was run 10 times for each precipitation perturbation and each temperature perturbation.

Using only 10 of the 50-year stochastic time series of resampled historical values, and choosing only four-fifths of the length of each time series (because the WEAP model is set up to receive only 40-year time series), time series of future climates were generated. First, a decision was made about the ranges over which to perform the stress test. The relevant downscaled GCMs range from approximately 650 to 1,325 mm of precipitation in 2040, and average annual temperatures of approximately 20.3 to 22.7°C (see figure 4.1). Understanding that the range of future climates indicated by the GCMs should serve as a minimum indicator of uncertainty, this analysis extended somewhat beyond the range of the GCMs. Thirteen "bins" of precipitation and six "bins" of temperature were used. The binning of precipitation and temperature conditions into categories of plausible climate trends was done by placing a multiplicative trend on precipitation, to produce annual average precipitation totals for 2040 in the range of 400 to 1,600 mm, or −61 to +57 percent of historical, in increments of 100 mm; an additive trend was placed on temperature, to produce changes in annual average temperature for 2040 of −1, 0, +1, +2, +3, and +4. The ultimate temperature and precipitation increases were achieved in 2050, having begun in 2010, meaning that each year in between received one-fortieth of the total change. Figure 4.1 and all subsequent figures show the average of GCM projections from 2030 to 2050, centered on 2040.

In total, for the prefeasibility design, the WEAP model was run for 780 climate states (13 precipitation, 6 temperature, 10 runs of each). The WEAP model calculated a time series of monthly streamflow at each site, as well as the monthly hydropower generated. The levelized cost and levelized cost regret were calculated during post-processing of the WEAP results.

**Mean Climate Response Surface**

Refer again to the climate response surface, introduced in chapter 3 and figure 3.9. If, for example, attention were given to the prefeasibility design (31 m³/s), and a performance threshold (in this case, levelized cost regret) of $500/GWh had been set, the climate response map would identify climate states of unacceptable performance relative to the threshold. In the red area (low precipitation and relatively high temperature), the levelized cost regret of hydropower generated by the prefeasibility design is higher than $500/GWh, which means the design is a failure. However, the system performs acceptably in the green region. Scenarios to be further evaluated, therefore, lie in the red region and in the part of the green region near the failure threshold.

The system is obviously sensitive to climate (Phases 1 and 2). The summary findings of Phase 3, illustrated in figure 3.9, demonstrate that, more than being generally sensitive, this particular design has substantial climate vulnerabilities within the range of climate change explored, and, significantly, within the range suggested by both downscaled GCMs and by extrapolation of the historical trend. *This is the first input received from the GCMs in this analysis. The role of the GCMs has not been to project system performance, but to provide information about the relative likelihood of poor system performance, after the vulnerability assessment was completed using projected climate time series generated by the stochastic weather generator.* Assuming that the decision maker, taking stock of the GCM projections and historical trend (and any available paleoclimatology data not shown here) is dissatisfied with the magnitude of the climate-related vulnerability of the proposed system, design modifications (or alternative projects altogether) would be explored in Phase 4.

## Phase 4: Climate Risk Management

At this point, significant vulnerabilities in the hypothetical prefeasibility design have been identified. An initial step toward the management and reduction of those climate-related risks might be to test other turbine capacities. This section describes a simplified version of the process used by decision scaling to manage climate risks. Because this example is simple, most of the advanced search algorithms (for example, real options analysis, robust optimization) and scenario-clustering procedures (for example, robust decision making) for decision making under uncertainty are unnecessary. Were the problem more complex, the more advanced (and more time- and computationally intensive) techniques might be warranted. These methods are briefly discussed in chapter 5.

### Return to Phase 3: Mean Climate Response Surface

There is no reason to believe that robustness in this case is not achievable. Phase 4 therefore involves strategic design modifications, with each revised project design configuration being resubjected to the Phase 3 Climate Stress Test. Figure 3.9 demonstrated the climate response of the prefeasibility design, thereby identifying its mean climate-related vulnerabilities. Figure 4.2 presents the results of a climate stress test on a range of 12 design capacities (of 29, 31, 33, 35, 37, 39, 41, 43, 45, 47, 49, and 51 m³/s), and identifies the band of climate space over which each design operates with the lowest cost regret. Costs are calculated using a pricing spreadsheet in which the cost of each design is discreet and nonlinear relative to the cost of each other design.

To perform this analysis, a Visual Basic for Applications (VBA) script was used to call and run WEAP (see brief description of WEAP in appendix A) for every climate time series for every design. The WEAP model was run for the 12 designs and 780 climate folders, resulting in a total of 9,360 WEAP runs. Each time WEAP was run, the VBA script provided it with a design capacity (North tunnel size, South tunnel size, and turbine size) and the appropriate number of the climate folder containing the climate time series. The WEAP model then calculated a 40-year time series of monthly streamflow at each site, as well as the monthly hydropower generated. Results were output to folders to be analyzed in R (an open-source software environment for statistical computing and graphics).

Figure 4.2 is a combined climate response map that enables the performance of the 12 alternative designs to be compared (design 1 having the smallest capacity and design 12 the largest). The dashed domain is the climate envelope for the CMIP3 ensemble, the dotted domain is the climate envelope for the CMIP5 ensemble, and the solid box encapsulates the climate envelope for all projections plus the historical observations (see "Risk Analysis Concepts" in this chapter for a discussion of the rationale behind the development of the three climate envelopes). In brief, polygons can be drawn to

**FIGURE 4.2  Climate Stress Test Results**

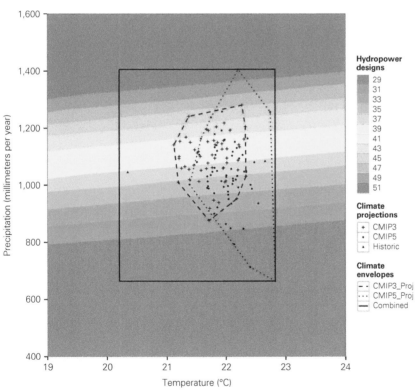

*Note:* CMIP = Coupled Model Intercomparison Project. The values of the general circulation models presented are the average of the time series from 2030 to 2050. The axes are "bins" that represent the mean of 10 stochastic time series, based on changes in historical means by 2040. For example, the red area at 600 mm and 19°C represents the preference for (lowest levelized cost of) design 1 (29 m³/s), subjected to a 40-year (2010–50) time series of climate data with average precipitation of 600 mm and temperature of 19°C between 2030 to 2050.

encapsulate the area on the climate response surface in which GCMs project future temperature and precipitation, and the area can be assigned a uniform distribution (giving no preference to GCM "clusters"). If the resulting encapsulated space in the climate response surface is all red (regardless of shade of red), the chance of unacceptable system performance is estimated to be 100 percent; adaptation is thus necessary.[2] If the space is only one-quarter red, the likelihood of unacceptable system performance is 25 percent, and risks (for example, net present value, discounted costs) should be calculated to determine whether adaptation measures are worthwhile. In the current example, if average annual precipitation exceeds about 1,300 mm, the preferred design would be the largest design (design 12, at 51 m³/s), to be able to make productive use of the available streamflow. However, if average annual precipitation is less than about 850 mm, the smallest hydropower design capacity (design 1, at 29 m³/s) would be the most efficient. "Most efficient," in this case, means that the design under consideration would operate at the lowest levelized cost. Temperature also affects the choice, but it does so to a lesser extent than precipitation; at higher temperatures, more precipitation is required to warrant the choice of a larger design.

### Digging Deeper into Levelized Cost Regret

Figure 4.3 presents levelized cost regret for four of the alternative hydropower design capacities. As demonstrated by the white area in panel 1 of figure 4.3, design 1 (29 m³/s) is the preferred design when future mean precipitation is less than about 1,000 mm. In the white area, the levelized cost regret is zero, which means that of all the considered alternatives, design 1 has the lowest levelized cost regret within that particular climate space. Design 12 (51 m³/s capacity, panel 4 of figure 4.3) produces hydropower with the lowest levelized cost for future precipitation values higher than about 1,100 mm in 2040 (with linearly increasing precipitation trends between now and then). *For the entire modeled climate domain, larger designs have higher potential for higher levelized cost regrets than do smaller designs*, keeping in mind that no GCMs project conditions in which the larger designs have their highest levelized cost regret (these lie outside the solid box), and only two extreme GCM projections indicate the likelihood of even "high" levelized cost regrets for large design capacities (at the lower right corner of the solid box). Looking specifically at the CMIP3 envelope, the CMIP5 envelope, or the region of extrapolation of the historical trends, however, may result in different conclusions. Further exploration is needed, and is provided in the "Risk Analysis Concepts" section.

FIGURE 4.3 **Levelized Cost Regret**

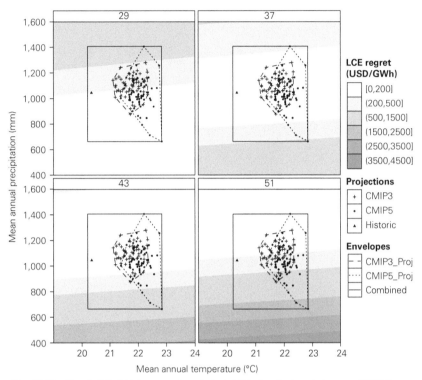

*Note:* CMIP = Coupled Model Intercomparison Project; LCE = levelized cost of energy ($/GWh), as calculated in equation 4.1; m³/s = cubic meters per second.

Turning briefly to an evaluation of total net benefits, figure 4.4 shows that, in general, larger designs potentially yield higher total net benefits. This is an important finding in favor of larger designs, given that figure 4.3 indicates that the levelized cost regret metric favors smaller designs. Within the envelope of projected and historical climate (solid box), at a price of $0.05/kWh (best current estimate of the future local price of electricity), design 12 produces the largest total net benefit (zero net benefit regret, that is, forgone profit, throughout). Design 1, however, could potentially result in more than $200 million of forgone profit relative to the highest-profit-generating design within the domain of projected and historical climate. Figure 4.4 is, of course, highly sensitive to the forecasted future price of electricity, and a sensitivity analysis should therefore be performed on this factor.

FIGURE 4.4 **Net Benefit Regret**

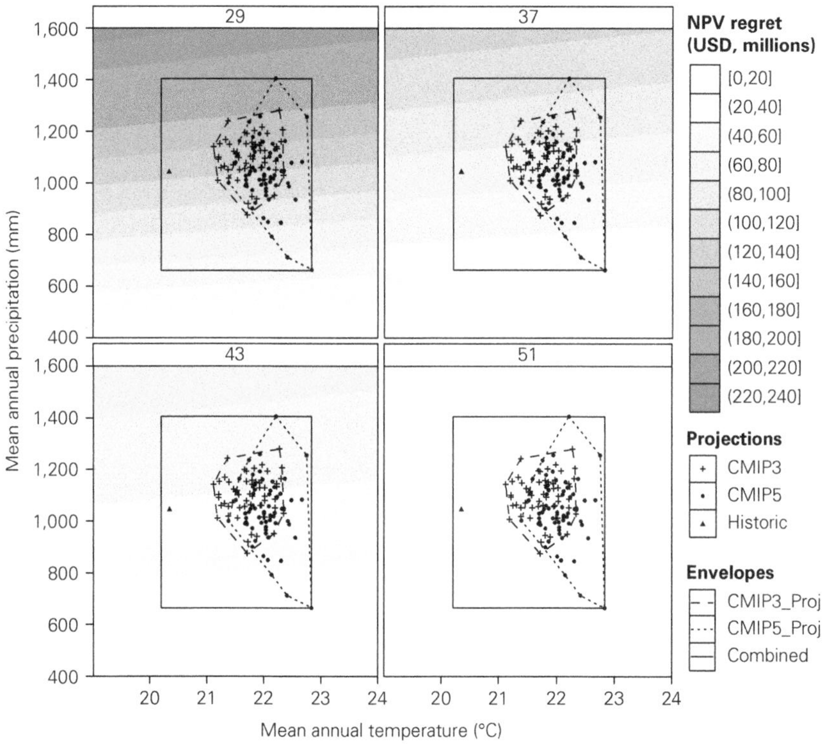

*Note:* CMIP = Coupled Model Intercomparison Project; NPV = net present value.


## Risk Analysis Concepts

Figures 4.5 and 4.6 summarize the findings of a straightforward analysis of the risks faced by various project designs. The performance of each design is presented in every panel as measured by (1) the considered futures over which it has a levelized cost regret of zero (the lowest levelized cost among all alternatives)—a measure of robustness, (2) its expected levelized cost regret ($/GWh)—a measure of robust optimality, and (3) its total net benefits ($/year)—an alternative measure of robust optimality. Each figure further subdivides the results specific to the performance of each design in the domain of CMIP3-projected space, the domain of CMIP5-projected space, and the total climate envelope (designated in figures 4.5 and 4.6 as "Full range"). The ranges of performance of each design across the 10 stochastically generated future climate realizations are presented as colored lines above and below mean performance (black dots).

As shown in equation (4.2), the basic concept of risk, applicable in this and most other instances of water resources planning and management, is a combination of impact (hazard) and the probability of that impact.

$$Risk = \sum_{i=1}^{\text{all future states}} Prob_i \times Impact_i \qquad (4.2)$$

The probabilities in figure 4.5 are derived from GCM counts, as was demonstrated for the prefeasibility design in the "Phase 3: Climate Stress Test" section of chapter 3, and a uniform distribution is used to characterize the likelihood of each GCM result. A count of the downscaled GCMs falling in the region of preference for each design is not the only (or even the preferred) method for evaluating the relative likelihood of each climate state (and thereby the relative likelihood of the preference for each design). Were the GCMs independent, this might be a reasonable approach. However, "clustering" of GCMs in particular regions of the climate response map may

FIGURE 4.5 **General Circulation Model Count–Based Relative Preferability of Designs**

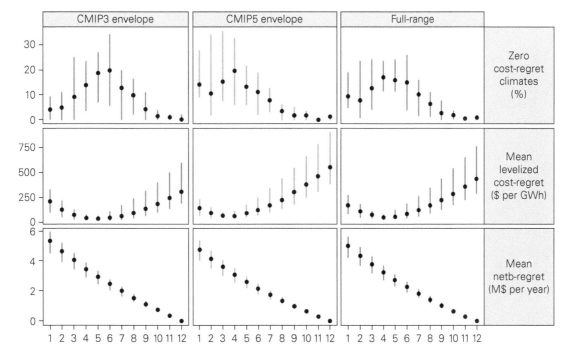

*Note:* CMIP = Coupled Model Intercomparison Project; GWh = gigawatt hour; M = million; netb = net benefit. Mean levelized cost regret is calculated using equation 4.1.

be attributable to the cross-model similarities in GCM model design and the propagation of potentially erroneous common assumptions (or data sources) throughout numerous GCMs. In this example, only 23 separate GCMs were run, using a number of different CMIP3 (A1B, A2, B1) and CMIP5 (RCP 4.5, 8.5) emission scenarios. Figure 4.6 therefore presents the results of an alternative approach that assigns a uniform[3] distribution to the region of the total climate envelope and weights each alternative design by the fraction of the total climate envelope in which it is preferred.

According to figure 4.6, larger design capacities are likely to have the highest levelized cost regret and the lowest net benefit regret. Designs in the low-middle range have zero cost regret for the largest number of GCM projections, and also the lowest mean cost regret. Less uncertainty (shorter uncertainty bars) surrounds net benefit regret projections than levelized cost projections (assuming the forecasted electricity price is reliable).

The key finding of figure 4.6 is that design 1 is highly cost-efficient over a large portion of the climate envelope. However, because the region of the climate envelope in which design 1 is highly cost-efficient contains few GCM

**FIGURE 4.6  Climate Envelope–Based Relative Preferability of Designs**

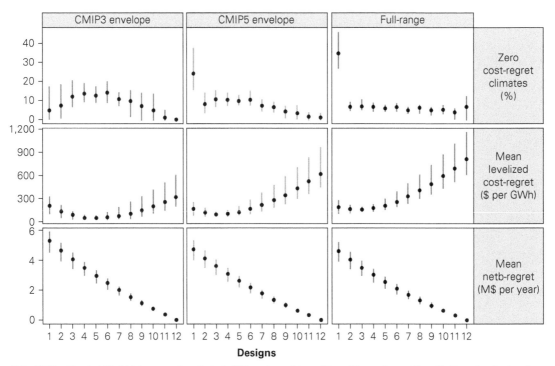

*Note:* CMIP = Coupled Model Intercomparison Project; GWh = gigawatt hour; M = million; netb = net benefit. Mean levelized cost regret is calculated using equation 4.1.

projections, it does not show up so prominently in figure 4.5. Interestingly, design 1, while maximizing the probability of zero cost regret in figure 4.6, does not produce hydropower with the lowest expected levelized cost regret because it has very high cost regret in the small regions of the climate envelope in which it has any cost regret. In the risk matrix (table 3.1), high regret in design 1 is a high impact–low probability event, and equation (4.2) (the mathematical expression of risk) shows that the risk of that event actually occurring (the product of high impact and low probability) is small. This close examination of design 1 in figure 4.6 is therefore a useful exercise in understanding how the source of climate information could affect the design. A few anomalous climate projections in the CMIP5 ensemble point to the preferability of design 1. However, on the whole, design 1 is not very robust. The design does command attention in the CMIP3 group.

## Discussion and Recommendations

If minimization of maximum net regret is the driving motivation (conservative decision making), smaller designs would be preferred. If maximization of net benefits is the goal (opportunistic decision making), larger designs would be preferred. Design 12 looks especially robust for decision makers driven by profit maximization. Of course, to achieve greater future returns, a bigger investment must be made in the present to build larger hydropower facilities. Budget constraints should therefore be considered.

The results make a strong case for designs in the middle range, which have relatively low levelized cost regret and relatively low net benefit regret. Such middle-range designs are robust to most projected future climate states. Design 1 produces hydropower with the maximum probability of zero levelized cost over a large portion of the climate envelope; however, it is not a portion of the space in which many GCM projections fall. Given the competitiveness of several of the evaluated designs, other metrics, such as capital investment required, investment resources available, and impact on existing ecological and human systems, should be used to distinguish between the choices. Along these lines, in an adaptive framework, it would be ideal if the hydropower system could be designed to be modular and expandable with marginal additional investment in the future, if higher precipitation occurs.

Furthermore, this simple case study presents design sensitivities to changes in mean annual precipitation and mean annual temperature. The hypothetical hydropower system has been shown to have sensitivities to changes in precipitation variability. A second climate stress test could therefore be developed to evaluate the relative performance of each alternative design across a range of streamflow means and variabilities (for example, increases in the intensity of precipitation resulting from a narrowing of the monsoon season). The ability of weather generators such as the one developed for this study to perturb many aspects of the climate signal—mean values (annual, monthly, or daily precipitation or temperature), extreme values, variability, and spatial and temporal patterns, and the like—is a particular strength of this approach. The insights gained into the broader system sensitivities through a more comprehensive analysis of climate parameters, in addition to annual average temperature and precipitation, could be instrumental in the identification of preferred system designs, especially when placed in the context of best scientific understanding of likely changes in seasonality, variability, and other spatial and temporal patterns. Once a weather generator has been developed, the additional computational effort involved in such explorations is not prohibitive.

Finally, returning to the statement in the introduction to this chapter on the use in this example of unrealistically small thresholds for cost regret, the implications for the application of the decision tree to this example are clear: this example deserves a more thorough exploration in Phase 2. Were other uncertainties considered (for example, demographic factors such as population growth and per capita use of water and electricity, agricultural water use, electricity price, and discount rate), they would likely be found to be relatively more significant to project performance than climate uncertainties. For the sake of illustration, this analysis jumped straight to an examination of the effects of climate-related uncertainties. A more thorough Phase 2 evaluation would begin with a quantification of the elasticity of the prefeasibility design's levelized cost (or net benefits) to changes in nonclimate uncertainties. If the elasticity is significantly higher for nonclimate factors, a Phase 3 Climate Stress Test may not be necessary. Instead, attempts to reduce the uncertainty in nonclimate factors would be made in order of greatest to least levelized-cost or -benefit elasticity. Adaptations may be made to the prefeasibility design to address potential nonclimate vulnerabilities, with climate-related vulnerabilities addressed only if significant potential vulnerabilities remain.

## Notes

1. Special thanks to Mehmet Umit Taner for his contribution in the analysis and writing of this chapter. His manuscript detailing this case study is forthcoming. The reader may recognize in this example the configuration of the Lower Fufu Hydropower Project in Malawi (Cervigni et al. 2015). The example has been adapted and presented here only to illustrate the application of the decision tree

procedure. Conclusions refer to the hypothetical example only and cannot be used to indicate the relative merits of the actual proposed project.

2. The shade of red indicates the severity of the failure, and is useful as an indication of the magnitude of vulnerability, but is not immediately relevant to the binary question of failure or not.

3. It is not necessary that the distribution be uniform. Other distributions, such as bivariate normal, are equally possible. However, the use of the uniform distribution is simplest and quickest, and requires the least evidence of applicability. If a bivariate normal distribution were assigned to the region of the total climate envelope, each alternative design would be weighted by the fraction of the total envelope in which it is preferred, and the quantile of the bivariate normal distribution in the region of preference.

## References

Cervigni, R., R. Liden, J. E. Neumann, and K. M. Strzepek. 2015. *Enhancing the Climate Resilience of Africa's Infrastructure: The Power and Water Sectors.* Washington, DC: World Bank.

Lall, U., and A. Sharma. 1996. "A Nearest Neighbor Bootstrap for Resampling Hydrologic Time Series." *Water Resources Research* 32 (3): 679–93.

**CHAPTER 5**

# Further Guidance for Decision Making under Uncertainty

## Introduction

The methodology presented in chapter 3 describes a decision tree framework for bottom-up, climate-informed decision making in water resources planning and management. It guides project planners through a risk assessment (tailored to climate risk, but generalizable to most other risks a water system might face) in which analytical effort accounting for the uncertain effects of change in some conditions is expended in proportion to the project's sensitivity to those particular conditions. If the assessed risks to the project are significant, tools for decision making under uncertainty may be needed to systematically modify the proposed project design to reduce its vulnerabilities. This chapter provides background on prominent examples of such tools.

## Background

Reports on decision making under risk and uncertainty in water resources planning and management have proliferated recently. Most of these reports have been generated at the request of governments, international agencies, or donors (Coates et al. 2012; Hallegatte et al. 2012; National Research Council 2009; Ranger et al. 2010; Schultz et al. 2010; Vucetic and Simonovic 2011). The reports together represent a compendium of best practices for planning water resources projects in the absence of a clear picture of future conditions. All of the reports emphasize the fundamental role of uncertainty in water resources systems decision making in light of the increasingly significant nonstationarity of the climate signal. Most of the reports make clear that water systems have always been planned under uncertainty, and that the nonclimate uncertainty facing water systems planners has often outweighed the uncertainty related to future climate. Therefore, the methods pointed to for decision making under this new, "deeper" or "more severe" climate uncertainty (as discussed in the following section) are mostly drawn from methods previously developed for incorporation of more conventional uncertainty in water systems planning.

This chapter summarizes the findings of those reports and organizes the information to correspond with the phases of the decision tree (see figure 3.1). A number of concepts and techniques are introduced that are explained in greater detail in later sections. Note that not all of the methods described herein are exclusively decision-making approaches; some include tools for risk assessment that can be incorporated into Phases 1–3.

### Extensions to Traditional Decision Analysis

Operations research, developed during World War II, has provided tools for modern decision analysis. The foundation of operations research is the formulation of a mathematical model to represent a problem, and the use of a computer-based procedure for deriving solutions to the problem (Hillier and Lieberman 2005). Traditional decision analysis aims for an optimum system configuration (for example, infrastructure development or operational policy), typically by minimizing expected cost or maximizing expected benefits.

Some tools for traditional decision analysis are built on search (optimization) algorithms (Kasprzyk et al. 2013; Loucks 1970; Ray, Kirshen, and Watkins 2012; Steinschneider and Brown 2012). Others run simulations of the system repeatedly (a so-called Monte Carlo experiment), systematically varying model parameters to identify preferred system configurations (Jeuland and Whittington 2014; Lempert et al. 2006; Lempert and Groves 2010; Prudhomme et al. 2010). If a tool based on a search algorithm is used, emphasis should be placed on the generation of trade-offs between competing objectives (Kasprzyk et al. 2013; Ray et al. 2014). Robustness with respect to one objective (for example, maintenance of low flows for ecological well-being) will likely be won at the expense of some other objective (reservoir storage for water supply during drought, for instance). Other concepts to be emphasized in optimization objective functions are adaptability (real options analysis [Ranger et al. 2010]) and diversification (redundancy and diffusion of risks [Brown and Carriquiry 2007]). A well-crafted optimization model for water systems planning and management will likely incorporate elements of all three objectives—adaptability, diversification, and robustness.

Traditional decision analysis tools and their extensions are suitable for the types of projects that exit from Phase 2 or Phase 3 of the decision tree, but are of limited utility to problems involving "deep" or "severe" uncertainty, as would be analyzed in Phase 4. For projects exiting at Phase 3, in particular, the cost-benefit analyses performed must pay careful attention to the inclusion of safety margins and sensitivity analysis, given that those projects were shown in Phase 2 to have significant potential sensitivities to climate change, though current climate change projections do not indicate a high likelihood of resulting system failure (relative to performance threshold). Phase 3 problems might also benefit from treatment using models based on stochastic optimization procedures, as discussed later in this chapter. The tools needed for Phase 4 analysis are summarized in the "Risk-Management Tools" section of this chapter, following an introduction to more traditional techniques and a short discussion of key concepts.

### Trade-Offs between Benefits and Costs

In water resources applications, a system model tends to take water-related parameters as input and outputs economic metrics. A water supply system model might allocate the water according to minimum cost or maximum utility. A water distribution network (or wastewater collection network) system model would provide guidance on the relative costs and benefits of reconstruction or reoperation[1] of the storage, treatment, and channel system. A flood control system model would likely seek to minimize the cost of flood damage. In each case, in addition to parameters related to the current system configuration, historical performance, and comprehensive cost and benefit data, water system models used as planning tools need information that anticipates the behavior of water within the system during the lifetime of the project. Hydrologic models translate forecasts of temperature and precipitation into water runoff or water storage. A description of a number of the hydrologic models most commonly used in the United States is included in appendix A.

A decision-support model of the system can be developed to assist in the characterization and quantification of the economic (and other) trade-offs inherent in the project installation. Use of a hydrologic model to project water (runoff, soil moisture, storage, seepage, and so on) in the system is not entirely necessary, but many such tools are available, such as lumped parameter tools (for example, the abcd model) and distributed models (for example, variable infiltration capacity—the VIC model), summarized in appendix A. However, a model of the benefits and costs of each project is absolutely necessary to compare the relative merits of each project to its alternatives. Output of such a model could be monetary benefits, metrics of system performance (for example, reliability or invulnerability), or ecosystem costs. Because every water-related project, program, or policy entails costs, a framework is needed to permit comparison of the net benefits of the proposed project to its alternatives. The framework need not monetize all costs and benefits, and need not be overly complex or even computer based. However, without the establishment of a framework for trading off benefits and costs, decisions for investment in water projects lack basis.[2]

Conventionally, water resources system planners, faced with uncertain future benefits and costs (and their more fundamental elements like demand and supply), would use deterministic tools to optimize over expected values. Being aware that the use of *expected values* was subject to the potential realization of futures less fortuitous than the expected, planners attach *safety margins* to key system elements or operating policies. Safety margins, *sensitivity analysis* (as described below), and *adaptive management* (described later in this chapter both as a concept and in practice) are tools for adding robustness to traditional decision analysis. However, traditional expected-value decision making, even in combination with tools for adding robustness, has significant drawbacks when the sensitivities of the system are substantial (Loucks, Stedinger, and Haith 1981).

### Safety Margins

A simple and effective strategy for decision making under uncertainty is to be conservative, traditionally accomplished by including a safety margin (also called a safety factor). For example, if it is believed that sea level will rise approximately 1 foot in the next century, it would make sense to build a 2-foot sea wall to hedge against a worst-case scenario. *The more expensive the potential failure, the more reasonable it is to include a very large safety factor.* If the proposed sea wall is intended to protect New York City from a 1-foot sea level rise, it might be reasonable to include a factor of safety of 1,000 percent and build a 10-foot sea wall. Of course, the magnitude of the safety margin is affected by many factors, including the cost of additional capacity, the consequences of system failure, the economic lifetime of the project, the flexibility of the design, and the likelihood that better forecasts of future conditions will become available in time to add additional capacity at a later stage. The discussion of real options analysis (ROA) later in this chapter presents, as an example, a method for inclusion of extra strength in the base of a sea wall that would enable extension of the sea wall height as needed in the future.

### Sensitivity Analysis

For situations in which uncertainty exists about the system model or the distribution of its inputs, practitioners of traditional decision analysis turn to sensitivity analysis. Sensitivity analysis is a simple method for assessing the effect of uncertainty on system performance by considering the possible costs of making other than the optimal decision. According to Loucks and van Beek (2005, 261), "A sensitivity analysis attempts to determine the change in model output values that results from modest changes in model input values. A sensitivity analysis thus measures the change in the model output in a localized region of the space of inputs."

The shadow price in linear programming is a particularly useful tool in sensitivity analysis because it presents the marginal utility of relaxing a system constraint by one unit. In the run-of-the-river hydropower case study presented in chapter 4, for example, it would be possible to show the marginal value (as measured by a decrease in levelized cost) of 1 additional millimeter of rainfall per year. A sensitivity analysis is not the same as a thorough analysis of the uncertainties potentially affecting system performance (together with their probability of occurrence), nor does it address the question of what decision should be made when the future is unknown or unknowable (Loucks, Stedinger, and Haith 1981).

Saltelli, Tarantola, and Campolongo (2000) describe four primary contributions of sensitivity analysis:

- It addresses structural uncertainty and provides guidance for the identification of the weak links in a scientific assessment chain.
- It can be used to determine which subset of input factors (if any) accounts for most of the output variance (and in what percentage).
- It can be useful as a quality assurance tool, to make sure that the assumed dependence of the output on the input factors in the model makes physical sense and can be reconciled with the analyst's understanding of the system.
- It can be used before and during model identification and parameter estimation to calibrate the degree of detail of the model to the task at hand, saving the analyst time.

However, as argued by Lempert et al. (2006), the attachment of sensitivity analysis to traditional decision-analysis techniques is an adequate measure for risk exploration only if the optimum strategy is relatively insensitive to key assumptions. If it is not, sensitivity analysis techniques "can encourage analysts and decision makers to downplay uncertainty to make predictions more tractable" (Lempert et al. 2006, 515). Furthermore, they can lead to strategies vulnerable to surprises that might have been countered had available information been used differently (Lempert, Popper, and Bankes 2002). Thus, conclusions drawn from sensitivity analysis should be handled judiciously.

## Key Concepts in Decision Making under Uncertainty

Before proceeding to a discussion of the types of advanced tools for decision making under uncertainty that are useful in Phase 4, beyond the simple extensions to traditional decision analysis just presented, two key concepts that surface repeatedly in the literature on decision making under uncertainty should be defined: the concepts of "deep" or "severe" uncertainty, and of "robustness." These terms are ambiguous, so various perspectives on them are offered here. The community of academics and practitioners developing these ideas continues to assess the degree to which a formal definition of these concepts is necessary, and under what conditions.

### Deep and Severe Uncertainty

According to robust decision making (RDM), "deep uncertainty is the condition in which analysts do not know or the parties to a decision cannot agree upon (1) the appropriate models to describe interactions among a system's variables, (2) the probability distributions to represent uncertainty about key parameters in the models, or (3) how to value the desirability of alternative outcomes" (Lempert et al. 2006, 514). Information gap decision theory (IGDT) defines severe uncertainty as "conditions where the evidence upon which to base a decision is scarce and only of limited relevance to predicting what may happen in the future" (Hall et al. 2012, 4–5). Such uncertainty leads to an information gap—a disparity between what is known and what needs to be known to make a dependable decision.

For the purpose of decision making under uncertainty, the common thread between the two concepts is that they cannot be characterized by a single probability distribution. (See box 5.1.) In this way, both are examples

**BOX 5.1**

## Deep and Severe Uncertainty

*Deep uncertainty* refers to the condition in which probability distributions cannot be assigned to the key uncertainties, the appropriate models to describe interactions among a system's variables are lacking, the relative desirability of various alternative outcomes cannot be quantified, or any combination of the three.

*Severe uncertainty* refers to conditions in which an unbreachable disparity exists between what is known and what needs to be known to make a dependable decision.

For the purpose of decision making under uncertainty, the common thread between the two concepts is that they elude characterization by a single probability distribution.

of Knightian (1921) uncertainty. A common modeling response to deep or severe uncertainty is to characterize all events in the possibility space as equally likely, and to assign them a uniform distribution. This assignment does not solve the problem of needing to select bounds for the uniform distribution, and the net effect is that all future scenarios considered "possible" are assigned equal probability of $1/N$ (N = number of scenarios), whereas all future scenarios outside of the set explicitly considered are assigned a probability of zero (thus labeled impossible). Better solutions have been adopted by both the IGDT and RDM schools. The former does not use probabilities but rather considers nested sets of future states of the world representing the range of uncertainty around multiple parameters of interest. In this way, bounds on uncertainty can be usefully quantified. RDM can consider imprecise probabilities, represented as sets of alternative joint probability distributions over multiple parameters of interest, and then seeks strategies that are robust over a wide range of such distributions (Lempert and Collins 2007).

### The Concept of Robustness

Rosenhead describes robustness as a particular perspective on flexibility that is concerned with "situations where an individual, group or organization needs to make commitments now under conditions of uncertainty, and where these decisions will be followed at intervals by other commitments" (Rosenhead 1989, 188). A robustness perspective focuses alternately between the present, in which decisions must be made with the best available information about future conditions, and a regular updating of best estimates of those future conditions. The robustness of any initial decision is defined as "the number of acceptable options at the planning horizon with which it is compatible, expressed as the ratio of the total number of acceptable options at the planning horizon" (Rosenhead 1989, 190).

RDM has typically defined robustness as performing reasonably well compared with the alternatives over a wide range of plausible futures (Lempert et al. 2006). A system fitting this definition might be described as being "reliable" over a wide range of plausible futures, or possibly, depending on context, as having relatively low vulnerability. Applications of RDM have also defined robustness as "trading some optimal performance for less sensitivity to broken assumptions." (Lempert and Collins 2007, 1017) Variants of both these robustness criteria have been used in a recent World Bank report on water supply infrastructure (World Bank 2014). In particular, this report examines and compares climate adaptation strategies using three robustness criteria within an RDM framework: minimize maximum regret, satisfice over a wide range of future conditions, and satisfice over a wide range of likelihoods for future conditions. In most situations these criteria yield similar adaptation strategies, but in some cases the strategies diverge. Depending on the application, RDM applies robustness criteria either directly to measures of system performance (such as reliability, cost, and environmental impact) or to "regret" calculated from one or more of these measures. *Regret is the difference between the performance of some strategy in a particular future and the performance of the best strategy in that future.*

IGDT defines robustness as "the maximum uncertainty, measured by the parameter $\alpha$: $\alpha \geq 0$, over which a strategy achieves a certain level of performance... Robust-satisficing seeks to identify acts that perform acceptably well under a wide range of conditions... Robustness decreases as the requirement for reward becomes increasingly demanding" (Hall et al. 2012, 6–7).

Ray et al. (2014) explain that robustness, as with optimality or any other decision criterion, must be expressed with respect to some performance metric. A system cannot be said to be robust, generally, because a system cannot be said to be optimal generally. A water supply system that reduces water shortages by increasing spot market water transfers should not be described as "robust"; instead, it should be described as *robust to water shortages* (but certainly not robust to water *transfers*, which might have numerous undesirable qualities of their own). Specific to water shortages, for example, a robustness of 0.25 would mean that, based on the chosen water system capacity and the previous climate- and demand-related probabilities, there is a 25 percent chance that there will be a shortage in the modeled year. The system is thus robust to 75 percent of futures. However, this does not mean that it is robust to 75 percent of future scenarios, because the future scenarios are not necessarily all equally likely. But, based on the probability space, one would expect a 75 percent chance that the system will not fail (that is, a shortage will not occur).

When robustness concepts are applied to water systems decision models, precise language should be used. In the context of climate change risk screening, for example, water systems can be categorized according to (among others) each of four descriptors: (1) climate sensitive, that is, whether its performance is affected by climate at all; (2) reliable over a wide range of climate risks, that is, though it might be sensitive to climate change, its performance thresholds might not be violated; (3) vulnerable to very costly failures, that is, though it might resist failure, when it does fail, it might fail catastrophically; and (4) resilient, that is, able to recover quickly from failure to previous levels of performance. (See box 5.2.) Robustness can be presented with respect not only to system reliability in the face of flood or drought (as is typically done), but also with respect to a variety of occurrences of concern to stakeholders (for example, water transfers in response to drought or controlled flooding of farmland in response to a flood).

---

### BOX 5.2
## Robustness and Adaptability or Flexibility

Robust project designs perform reasonably well compared with alternative designs over a wide range of plausible futures. Robustness is a criterion that can be used to compare alternative decisions. One way to express the robustness of any decision is as a function of the number of possible futures (or size of the projected future domain) with which it is compatible, divided by the total number of projected futures (or total size of the projected future domain). A system fitting this definition might be described as being "reliable" over a wide range of plausible futures, or possibly, depending on context, as having relatively low vulnerability.

Adaptability or flexibility is an attribute of a decision that can be used to make a strategy more robust (or optimal). Adaptability or flexibility generally has costs, and so is preferable if (1) uncertainty is dynamic—knowledge is expected to improve over time; and (2) the project involves irreversible creation or destruction of capabilities. Certain adaptation strategies are more flexible than others to the possibility of upgrade in the future in the event that the impacts of climate change are high. Real options analysis is an established process by which adaptability can be explicitly incorporated into project designs through the use of Bayesian probability trees.

### Alternative Decision Rules

In traditional decision analysis, costs are minimized or benefits maximized. However, there are simple ways to represent a decision maker's degree of aversion to risk within the objective function. More-risk-averse decision makers might maximize the minimum profit or minimize the maximum regret. An objective function formulated to minimize the maximum regret would select the system configuration that results in the least bad worst-case performance (relative to other system configurations). The minimax regret formulation is the one used most often by those seeking a project robust to climate change, although methods like RDM and IGDT often use other robustness criteria, as described above.

Less-risk-averse decision makers, more interested in maximizing potential gains (for example, windfalls), might structure the objective function as a maximization of maximum profit. The optimization function in this case would seek the system configuration that would result in the occurrence of the maximum benefit achieved in any single scenario, regardless of the adequacy of that system configuration in other less fortuitous scenarios.

The Hurwicz criterion is a compromise between the maximax and maximin criteria. It uses a weighting method for balancing pessimism with optimism (Revelle, Whitlatch, and Wright 2004).

## Risk Assessment Tools

The following subsections summarize procedures for climate risk assessment—the scenario-neutral approach and IGDT—that are similar in concept to the stress-test elements of decision scaling, as presented in Phase 3 of the decision tree.

### Scenario-Neutral Approach

Prudhomme et al. (2010) perform a procedure very similar to the climate stress test described in relation to decision scaling. The authors describe the procedure as the "scenario-neutral" approach to climate change impact studies, and present it in contrast to "conventional, 'top-down' (scenario-led) approaches to climate change adaptation" (Prudhomme et al. 2010, 199). Both precipitation and temperature sensitivity tests sample from a range of scenarios significantly larger than those indicated by the projections. The study uses a "change factor" to apply an absolute percentage change to temperature and precipitation according to that suggested by the general circulation models (GCMs), and then uses a harmonic function to model the seasonal pattern of precipitation and temperature. The authors explain that "this is to allow for any significant difference in future projections from the next generation of climate models or new emission pathways to be part of the sensitivity domain" (Prudhomme et al. 2010, 206). By performing repeated simulations using a hydrologic model to observe flood peaks across scenarios, the authors gain valuable information (risk analysis) on the critical climate conditions at which the flood control system fails. The scenario-neutral approach does not present a method by which the decision maker can judge the relative merits of flood prevention strategies (as measured, for example, by cost). Robustness is characterized strictly as reliability.

## Information Gap Decision Theory

Hipel and Ben Haim (1999) argue that three fundamentally distinct approaches are available for formally describing uncertainty: probability, fuzzy set theory, and information gap modeling. Though there are ample examples of probability (the accepted standard and basis for most texts, see Revelle, Whitlatch, and Wright 2004; Loucks, Stedinger, and Haith 1981; Loucks and van Beek 2005; and Sen and Higle 1999) and fuzzy sets (Karmakar and Mujumdar 2006; Li, Huang, and Nie 2007; Li et al. 2009; Maqsood, Huang, and Yeomans 2005; Qin et al. 2007; Wang et al. 2009) used to describe uncertainty in water resources problems, IGDT modeling had not until recently been used in water resources. Since 2009, however, a number of studies have applied IGDT to a variety of water resources planning problems under great uncertainty, especially in flood mitigation (Hine and Hall 2010; Korteling, Dessai, and Kapelan 2013; Manning et al. 2009; Matrosov, Woods, and Harou 2013; Woods, Matrosov, and Harou 2011).

IGDT (Ben-Haim 2006) consists of three elements: (1) a nonprobabilistic, quantified model of uncertainty; (2) a system model that projects the outcome of decisions contingent on the model of uncertainty; and (3) a set of performance requirements that specify the value of the outcomes the decision makers require or aspire to achieve. The technique asks decision makers to set minimum and aspirational performance levels, and to favor the strategies that meet these levels over, respectively, the widest and narrowest range of uncertainty.

IGDT characterizes the uncertainty of system performance as a group of nested sets. The method requires the user to identify a best estimate of each unknown parameter from which to start the uncertainty analysis. Next, each of the input parameters is bounded in an interval, the range of which is meant to encompass "most" of the uncertainty particular to that parameter.

Whereas in the stress tests developed within decision scaling and scenario-neutral modeling a single increment of uncertainty is explored (the total range of average annual temperature and precipitation over which the performance of a water project is evaluated, for example), IGDT explores the range of performance within subsets of the total uncertainty space, which it refers to as "horizons."

IGDT scales the difference between the user-defined best estimate of a future parameter, $u$, and some particular realization, $\tilde{u}$. The deviation between $u$ and $\tilde{u}$ is scaled by $h$, which becomes the increment or uncertainty horizon. The IGDT uncertainty model is constructed as a nested set based on $\tilde{u}$ and $h$ such that $U(h, \tilde{u})$. A water system plan will have a range of performance for each uncertainty horizon specified by $U(h, \tilde{u})$. The minimum performance level is referred to as the *robustness*, and the maximum performance level is termed *opportuneness*. By quantifying opportuneness as well as robustness, the approach aims to be value-neutral—neither optimistic nor pessimistic, acknowledging that uncertainty means that outcomes could be both worse and better than expected.

IGDT requires the development of a system model and a nonprobabilistic quantified model of uncertainty. As with most other risk analysis tools presented here, the technique requires computing power to run thousands of simulations.

IGDT is in the early stages of establishing itself as a viable alternative to probabilistic conceptions of uncertainty. Given that probabilistic notions of deep and severe uncertainty are often ineffective, an effective alternative would be welcome. IGDT functions well in extreme uncertainty, when agreement cannot be reached on the probability distribution. It could be simply interval bounded, with all uncertain parameters nested and scaled by a common term $\alpha$.

Slowing its broader adoption throughout water resources planning and management are the difficulties presented to practitioners trained in probability theory who must learn IGDT concepts, which is a paradigm shift as an approach to uncertainty characterization. Effort must be expended to clearly communicate the model structure and findings, given that the technique requires a series of judgments from analysts and decision makers in constructing the uncertainty model and specifying the required minimum aspirational performance levels.

Another limitation of IGDT is that, in cases of great uncertainty, the user-defined best estimate of the parameter whose uncertainty is being explored may, in fact, be a poor estimate of the true value. In such cases, the largest nested uncertainty space drawn around the IGDT value might not be large enough to include the true value. In this sense, IGDT is not dissimilar from decision scaling and scenario-neutral modeling, both of which explore a limited space of potential future realizations of unknown parameters. Yet IGDT is particularly vulnerable to this shortcoming. Therefore, the "best estimate" of the uncertain parameter must be made with great care, and the uncertainty horizon explored should be drawn large enough to encompass all reasonable parameter realizations. Decision scaling improves on IGDT by starting from a logical point (climate normal) and then using projections to inform the probabilities of the space that can be derived.

This section has discussed IGDT as a risk-assessment paradigm. The aspects of IGDT that relate to decision making under uncertainty are discussed in the next section.

## Risk Management Tools

Traditional decision analysis and its extensions are ill-equipped to deal with the types of great uncertainty encountered in Phase 4 of the decision tree. This section therefore summarizes some of the more prominent tools available for decision making under deep (or severe) uncertainty. Decision scaling makes iterative modifications to the design to trace out alternative performance and find robust designs. IGDT and RDM take a similar approach, if by different procedures. Stochastic optimization, multi-(or many-) objective robust optimization, and ROA use search algorithms to quickly home in on the most efficient design modifications to improve system robustness.

### Information Gap Decision Theory

The risk assessment and uncertainty characterization aspects of IGDT have been discussed in relation to other tools for risk assessment. This section focuses on IGDT as a tool for decision making under uncertainty.

During risk assessment, the parameters of a candidate water plan are perturbed to explore the uncertainty space around a best estimate of future parameter values. The process is similar to that described for Phase 3 in chapter 3, but uses nested uncertainty increments instead of a single uncertainty space as illustrated in a climate response map. Both the robustness and the opportuneness of the candidate water plan are evaluated within each uncertainty increment.

Risk management, Phase 4 of the decision tree, begins with the identification of a number of other candidate strategies developed to reduce the vulnerabilities in the initial water plan revealed in Phase 3. If IGDT were used, the performance of each new system would be simulated, and the results of the simulations presented with robustness and opportuneness curves. In the instance of the run-of-the-river hydropower dam case study in chapter 4, the alternative water plans would be designs of different generating capacity. Previous studies have explored the relative robustness and opportuneness of different infrastructure portfolios for water supply (desalination, water import, and so on, as in Matrosov, Woods, and Harou [2013]), flood risk management options (for example, raising levee height or channel widening, as in Hine and Hall [2010]), or emission reduction paths (for example, Hall et al. [2012]), among others.

Robustness curves provide a count of the number of uncertainty increments in which the project being evaluated performs acceptably well. For the run-of-the-river hydropower development project, this might be a count of uncertainty increments in which levelized cost for each proposed design capacity is below some designated threshold. In an example from water resources planning, Matrosov, Woods, and Harou (2013) present robustness curves of storage susceptibility, total cost over the economic lifetime of the infrastructure portfolio, total energy consumption, supply reliability, and environmental factors. The larger the number of uncertainty horizons for which performance exceeds a given threshold, the more robust the infrastructure portfolio is to that threshold. Steep robustness gradients indicate minimal performance loss as uncertainty increases, and the points at which robustness curves cross mark the locations at which one infrastructure portfolio becomes relatively more robust than another.

Opportuneness curves show the potential windfall payoff for the optimistic planner. Riskier plans might have both higher potential robustness and higher potential opportuneness. The decision maker will want to weigh the potential for windfall in a riskier design against the consistent robustness of a more conservative design.

The decision maker using IGDT would evaluate the trade-offs between total system cost (or net benefits) and the robustness and opportuneness of the candidate water strategy and its various alternative designs with respect to all of the evaluated performance criteria. The ultimate decision, as is true for each tool for decision making under uncertainty, will be left to human judgment as informed by values, ethics, and risk tolerance.

## Robust Decision Making

RDM combines attributes of scenario planning, Monte Carlo simulation, and sophisticated analytical tools to identify vulnerabilities of proposed strategies, to help decision makers identify potential responses to those vulnerabilities, and then to identify which responses are most robust.[3] RDM is built on the premise that if adequate information were available about the future, one could conduct a traditional probabilistic decision analysis, which might seek to maximize net benefit or minimize cost. However, in the absence of high-confidence information about the relative likelihood of various possible futures, the most reasonable choice is to seek a strategy that satisfies one of the robustness criteria described in "The Concept of Robustness" above.

RDM inverts traditional sensitivity analysis, seeking management strategies (or project designs) whose good performance is insensitive to the most significant uncertainties. The process typically begins by running a simulation model multiple times with sets of input parameters that represent many different future states of the world. This formulation enables RDM to use many different types of climate information, including outputs from downscaled GCMs (top-down) (Groves et al. 2015), from stochastic weather generators that provide permutations based on historical climate records (bottom-up) (Groves et al. 2008), or from combinations of the two (Bureau of Reclamation 2012; Groves et al. 2013).

Regardless of how the climate scenarios are developed, the next step in the RDM process is to identify a candidate robust strategy by initially ranking or screening all proposed strategies according to a stakeholder-defined measure of performance, and to subsequently identify and characterize one or more clusters of future states in which each of the strategies performs poorly. These clusters, unweighted by probabilities, are designed to be considered unlikely or inconvenient even by decision makers. Faced with potential futures (or clusters of futures) in which a proposed strategy (or system) performs poorly, the RDM procedure makes modifications to the strategy (or system) that hedge against vulnerabilities. For the run-of-the-river hydropower project, once the risk assessment for the candidate strategy has been completed, RDM would simulate the performance of a number of alternative designs (for example, hydropower-generating capacities). Finally, trade-offs involved in the choice among the hedging options would be developed.

Because RDM samples from all combinations of uncertain system parameters, it explores futures both more benign and direr than the present. In a full RDM analysis, various aspects of candidate strategies (more than just the design capacity, as in the case of the run-of-the-river hydropower project) would be successively altered and resubmitted to the RDM process until a suitably robust strategy could be identified. A particular strength of RDM is its ability to consider a wide range of climate, socioeconomic, and other uncertainties, and to provide information on how system performance may be affected by combinations of such factors. The RDM framework makes it possible to analyze very large numbers of futures in which any or all of the system and design parameters are altered in any number of configurations.

Because of its dependence on modeling and computational complexity, RDM is computationally intensive. In Matrosov, Woods, and Harou's (2013) water resources system planning study, RDM implementation required approximately 310,000 simulations, whereas IGDT required only 1,600. Some RDM applications, however, have used as few as 200 futures. RDM incorporates scenario discovery algorithms, such as Friedman and Fisher's (1999) "patient" rule induction method, to identify and characterize clusters of future states in which the strategy fails to meet its goals, and requires many thousands of simulation iterations to be run to trace out system performance. The method has to be guided by an expert who understands the process of homing in from all possible strategies on those that are "robust" to their vulnerabilities and who can weigh the trade-offs between clusters of future world possibilities.

RDM is well developed and has been widely tested. The process presents the decision maker with trade-offs between robust strategies and identifies the vulnerabilities in every one of the proposed iterations' designs. It can consider a wide range of different types of uncertainties. For instance, it can identify water system vulnerabilities that arise from the interaction of climate and socioeconomic uncertainties. Such vulnerabilities may not appear when considering climate uncertainties in isolation.

RDM's primary weakness is that it can be computationally intensive. In some cases it can be useful to use a qualitative RDM analysis (McDaniels et al. 2012) or conduct an initial screening analysis to determine whether a full RDM analysis could prove valuable (World Bank 2014). When RDM is applied to climate-related uncertainty in particular, care should be taken not to rely too heavily on direct output from downscaled GCMs, and to sample a breadth of uncertainty beyond that suggested by the current generation of GCMs. In narrowing down the best-performing design alternatives, particular thought should be given to the conceptualization of all future states, extreme and otherwise, and especially whether each should be weighted equally in decision making.

## Dynamic Adaptive Policy Pathways (DAPP)

The DAPP approach[4] (Haasnoot et al. 2013) combines the concepts of adaptive policy making (Walker, Rahman, and Cave 2001; Kwakkel, Walker, and Marchau 2010) with those of adaptation tipping points (Kwadijk et al. 2010) and adaptation pathways (Haasnoot et al. 2011; Haasnoot et al. 2012). The key idea is to keep plans "yoked to an evolving knowledge base" (McCray, Oye, and Peterson 2010, 952). DAPP emerged in response to the desire among Dutch policy makers to make it easier to update plans in light of new climate scenarios (Kwadijk et al. 2010), and was adopted in the Dutch Delta Program. At the same time, in the United Kingdom, pathways were explored for the Thames Estuary 2100 study (Reeder and Ranger 2011).

Adaptation tipping points are a key concept of DAPP. An adaptation tipping point specifies the conditions under which a given plan will fail. This concept can be linked to other techniques such as RDM and decision scaling, which also focus on identifying the vulnerabilities of existing or proposed policies. An adaptation tipping point is reached when the magnitude of external change is such that the current management strategy no longer meets its objectives, and new actions are thus called for. The timing of an adaptation tipping point is scenario dependent.

The DAPP approach begins with the identification of objectives, constraints, and uncertainties relevant to a particular decision-making process. The uncertainties are next used to generate a variety of plausible futures. System performance in each of these futures is compared with the objectives to determine if problems are likely to arise or if opportunities are likely to occur. This analysis determines whether and when policy actions are needed, either through simulation-based techniques or optimization-based techniques. If simulation-based techniques are used, then a large number of model runs are performed, and the statistics of the results are analyzed to determine the general effectiveness of various design (or policy) alternatives. If optimization-based techniques are used, the best-performing suite of design alternatives and policies are identified by the optimization algorithm (and subject to the algorithm's particular biases and limitations).

To assemble a rich set of possible actions, the approach distinguishes between four types of actions: (1) shaping actions, (2) mitigating actions, (3) hedging actions, and (4) seizing actions (Kwakkel, Walker, and Marchau 2010). In subsequent steps, these actions are used as the basic building blocks for the assembly of adaptation pathways. The performance of each of the actions is assessed in light of the defined objectives to determine the adaptation tipping point of the action. Once a set of actions seems adequate, potential pathways (that is, a sequence of actions) may be constructed, and one or more preferred pathways can subsequently be selected as input to a dynamic, robust plan. The aim of this plan is to stay on the preferred pathway as long as possible. Contingency actions for staying on the pathway are defined, and a trigger for each contingency action is specified and subsequently monitored (Kwakkel, Walker, and Marchau 2010).

Figure 5.1 shows a stylized example of an adaptation pathway map, including a scorecard for each of the pathways. In the map, starting from the current situation, targets are first missed after four years. Following the grey line of the current plan, four alternatives become available. Based on an assessment of current projections of future conditions, Actions A and D are expected to be able to satisfy the performance threshold for the next 100 years *in all scenarios*. If Action B is chosen, a tipping point is reached in about another five years: the tipping point indicates that some performance metric will dip below the associated threshold in at least one future scenario. Achieving the set targets will require a shift to one of the other three actions defined (A, C, or D). If Action C is chosen, a shift to Action A, B, or D will be needed after approximately 85 years, the set targets would be achieved for the next 100 years, *except in the case of the worst-case scenario* (dashed green line). In the worst-case scenario, Action C would need to be abandoned after 85 years. The colors in the scorecard refer to the actions: A (red), B (yellow), C (green), and D (blue).

The adaptation pathway approach has several strengths. First, it is relatively easy to understand and explain. The metro map visualizations, as used in figure 5.1, are evocative, and help in visualizing the different routes that are available for achieving the desired objectives. Second, the approach

FIGURE 5.1 **Example of an Adaptation Pathways Map and Scorecard**

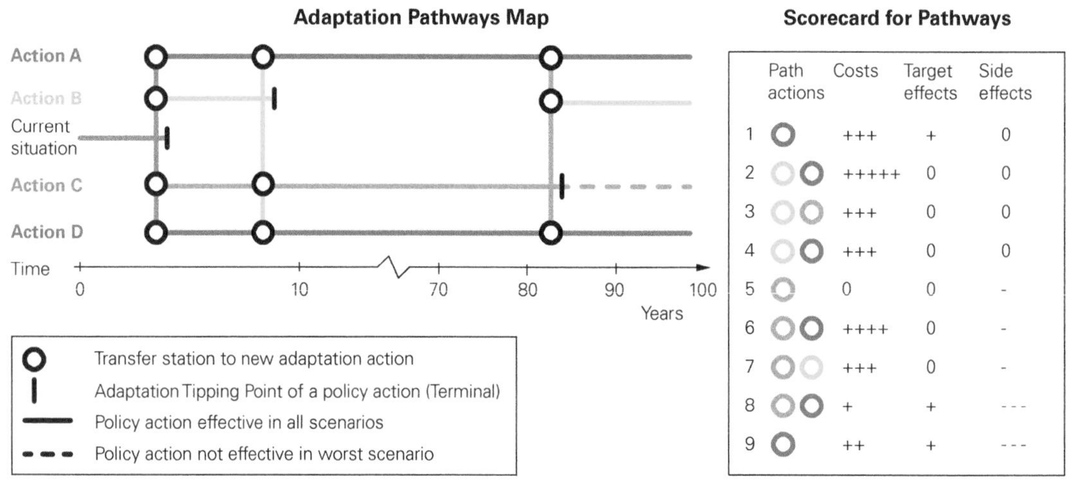

*Source:* Reproduced from Haasnoot et al. (2013).

*Note:* The map on the left shows the possible adaptation pathways, and the scorecard on the right presents the costs and benefits of the nine possible pathways depicted in the map. + = positive costs or effects (positive costs being "bad" and positive target or side-effects being "good"); - = negative costs or effects. More + or - presented sequentially indicates the more positive or negative the costs or effects.

Confronting Climate Uncertainty in Water Resources Planning and Project Design

encourages decision makers to think about "what if" situations and their outcomes, and to make decisions over time to adapt, while maintaining the flexibility to make future changes (Jeuken and Reeder 2011). This approach also helps designers anticipate undesirable lock-ins or other path dependencies, which can consequently be avoided. Third, the adaptation pathway approach explicitly frames adaptation as a dynamic process that takes place over time. It forces those involved to consider transient scenarios, rather than one or a few points in time (Haasnoot et al. 2012).

Although initially applied to managing climate change–related uncertainties and perspectives (Offermans, Haasnoot, and Valkering 2011), recent work has expanded the approach to cover other uncertainties as well, such as land use change (Kwakkel, Walker, and Marchau et al. 2014). The approach has also been put into actual practice. For instance, it is a cornerstone of the Dutch climate adaptation strategy in the water domain (van Rhee 2012). It was also a key element of the Thames Estuary 2100 study, and is being used in the development of climate adaptation strategies in New York (Rosenzweig et al. 2011; Rosenzweig and Solecki 2014), Bangladesh, Vietnam, and New Zealand (Lawrence and Manning 2012).

The adaptation pathway approach is continuing to be developed further. Related challenges include coping with multiple diverging objectives and a lack of clarity at present on exactly how to apply existing economic evaluation approaches to adaptation pathways. It is particularly difficult to account for the dynamic nature of adaptation pathways—the metro maps must be continually updated to reflect new information and new understandings of the anticipated best path forward. Similarly, most current examples of DAPP reflect a conservative, risk-averse, decision-making approach in which the switch is made from one pathway to another (with potentially significant transaction costs) when system performance in even a single future scenario is unsatisfactory in the first pathway (dashed lines in the "metro maps" notwithstanding), however dire and unlikely that future scenario may be. Improvements to the method could include probabilistic decision-making frameworks that are not dominated by the worst case. A particular difficulty in using adaptation pathways is that the metro maps, although visually highly instructive, can become overly complicated and cluttered for more complex problems. This complexity can be ameliorated somewhat by the relabeling and grouping of candidate actions.

## Stochastic Optimization

Stochastic optimization is an optimization technique in which uncertain future scenarios are weighted probabilistically and the resulting "best" design performs reasonably well across the range of considered futures. Stochastic optimization offers a first level of hedging against infeasibility, and thus is a step toward robustness. For summaries of stochastic optimization techniques that make use of probabilistic uncertainty paradigms for water systems decision making, see Revelle, Whitlatch, and Wright (2004), Loucks, Stedinger, and Haith (1981), Loucks and van Beek (2005), and Sen and Higle (1999). Multiobjective robust optimization (RO), briefly discussed earlier, extends stochastic optimization to make it more explicitly robust to challenging scenarios.

### Cost-Benefit Analysis under Uncertainty

Traditional decision analysis emphasizes optimality under an expected future. Cost-benefit analysis (CBA) is an example of this technique (Arrow and Fisher 1974). CBA under uncertainty (Arrow et al. 1996) is an improvement on CBA in the face of an unclear future, and is applicable to problems in which uncertainty is quantifiable. However, when representing the uncertainty associated with climate change indices (for example, temperature and precipitation) with Gaussian or other asymptotically diminishing probability distribution functions, the CBA under uncertainty method is extremely sensitive to tails of the distribution functions (Ray et al. 2014; Weitzman 2009). In situations of deep uncertainty, therefore, CBA under uncertainty is best used as a screening tool, and not as a replacement for the type of in-depth, bottom-up analysis adopted here (Hallegatte et al. 2012).

### Multiobjective Robust Optimization

According to Sahinidis (2004), there are three general methods for optimization under uncertainty: stochastic programming, fuzzy programming, and stochastic dynamic programming. Stochastic programming includes (1) the standard approaches using recourse models (termed two-stage or multi-stage stochastic linear and nonlinear programs; see Sen and Higle [1999] for an introductory tutorial on stochastic programming); (2) RO, as described below; and (3) probabilistic models (chance constraints, attributed to Charnes and Cooper [1959]); see Loucks, Stedinger, and Haith (1981) and Tung (1986) for an introduction to and applications of chance constraints to water resources problems. Potentially, a fourth class of methods for optimization under uncertainty—evolutionary optimization algorithms—can be connected to Monte Carlo simulation models of water resources systems (for example, Kasprzyk et al. 2009).

In contrast to chance constraint techniques, for example, which can only limit the probability of a violation of a model constraint, RO offers a means of simultaneously controlling the sensitivity of the solution to any uncertain parameters or inputs and penalizing exponentially larger violations of one or multiple model constraints. There are two forms of RO models: those that guarantee satisfaction of hard constraints, and those that apply penalties to violations of soft constraints. The field of manufacturing and engineering science has tended to emphasize formulations that guarantee the satisfaction of hard constraints, thus leading to single optimal robust solutions (Ben-Tal and Nemirovski 1998, 1999; Taguchi 1989). By contrast, Mulvey, Vanderbei, and Zenios (1995) recommend a mathematical programming approach based on a trade-off between solution robustness (nearness to optimality across all scenarios) and feasibility robustness (nearness to feasibility across all scenarios). Their RO formulation extends stochastic programming to a multiobjective optimization framework that includes higher moments of the objective value (variance, most commonly) and a penalty function or functions on violations of one or more chosen constraints. The scenarios used in RO are discrete points in an empirical probability distribution (or joint probability distribution), generated to represent best current understanding of the relative likelihood of potential future system states.

Applications of this type of RO in water resources range from water distribution system design (Cunha and Sousa 2010) and wastewater treatment design (Afonso and Cunha 2007) to the design of large-scale water systems (Escudero 2000), as well as the design of groundwater pump and treatment systems (Ricciardi, Pinder, and Karatzas 2009). Nearly all previous water resources applications involve only feasibility robustness, and do not consider solution robustness. Most such examples deal with groundwater remediation applications (for example, Alcolea et al. 2009; Bau and Mayer 2006; Bayer, Buerger, and Finkel 2008; Ko and Lee 2009; Ricciardi, Pinder, and Karatzas 2007). Only a few applications of RO to water resources systems also include solution robustness through the minimization of variance or standard deviation of direct cost in the objective function (for example, Kasprzyk et al. 2009; Kawachi and Maeda 2004; Ray et al. 2014; Suh and Lee 2002; Watkins and McKinney 1997).

### Real Options Analysis

As was pointed out in "Phase 4: Climate Risk Management" in chapter 3, a strong water system management plan will combine elements of adaptability and flexibility, diversification, and robustness. ROA is applicable when (1) uncertainty is more "dynamic" than "deep"—knowledge improves over time, and (2) the project involves irreversible creation or destruction of capabilities. Certain adaptation strategies are more flexible than others to the possibility of future upgrading if climate change impacts are high. The expected value of each option (more flexible and less) can be calculated and compared. The objective in this formulation is still to maximize net present value, but the adaptability of design options is explicitly considered. The government of the United Kingdom, for example, requires that climate change adaptation analyses account for "the value of flexibility in the structure of the activity" (HMT and Defra 2009, 14).

ROA is an established probabilistic decision process (and a subset of stochastic optimization) by which adaptability can be explicitly incorporated into project designs, and large potential regrets associated with either overinvestment or underinvestment in adaptation measures can be avoided. ROA encourages staged decision making, through which more expensive and more highly irreversible decisions are delayed until more information is available on which to base those decisions. The philosophical underpinning of ROA has roots in the work of Dewey (1927), who promoted policies that incorporate continual learning and adaptation in response to experience over time, and Rosenhead (1989), who presented "flexibility" as an indicator by which to evaluate the robustness of strategies under uncertainty. The mechanism for real options is founded in the analysis of financial decision making (Arrow and Fisher 1974; Henry 1974; Myers 1984; Copeland and Antikarov 2001).

ROA focuses on making changes to the system configuration in reaction to reductions to uncertainty through future learning (de Neufville 2003). Wang and de Neufville (2005) present real options analyses as subsets of real options "on" systems (focusing on the external factors of a system, with greatest benefit gained through financial valuation methods), and real options "in" systems (incorporating flexibility into the structural design). Most water resources engineering design problems are of the latter type.

The adaptive policy-making paradigm, of which ROA is one element, has received increased attention in water resources planning and management in recent years. As of 2013, dynamic adaptive plans were being developed or had already been developed for water management in New York, New Zealand, the Rhine delta, and the Thames estuary. In the Netherlands, ROA has been used to assess optimal costs and benefits of pathways for fresh water supply of the South-West Netherlands Delta, and for studying how flexibility can be built into flood risk infrastructure (Haasnoot et al. 2013).

Examples of real options for water supply include investments in pumps to draw upon dead storage, pipelines to connect to storage at another impoundment, or infrastructure to tap groundwater resources. Demand-oriented real options for water supply are also possible, such as investments in household metering and strong public outreach campaigns that could be activated at some cost to help enforce future conservation efforts (Steinschneider and Brown 2012). Real option water transfers provide a mechanism by which water supply can be augmented without the need for large-scale infrastructure expansion.

A number of studies have demonstrated how financial instruments such as leases, option contracts, and water banks can facilitate the trade of water between low- and high-priority uses in the event of a localized water shortage (for example, Brown and Carriquiry 2007; Characklis et al. 2006; Hansen, Kaplan, and Kroll 2014; Kirsch et al. 2009; Lund and Israel 1995; Palmer and Characklis 2009; Steinschneider and Brown 2012). Applications to water resources problems with a focus on the mitigation of flood damage have also become common (Gersonius et al. 2010, 2013; Hall and Harvey 2009; HMT and Defra 2009; Ingham, Ma, and Ulph 2007; Merz et al. 2010; Woodward et al. 2011; Woodward, Kapelan, and Gouldby 2014).

### Real Options Procedure

A net present value (NPV) method to evaluate and determine the most cost-effective options is preferable, but such a method would be flawed when future uncertainties are significant (especially if only a single future scenario is considered). "A 'managed adaptive' approach is used to track any changes in risk over time and manage these changes through multiple interventions, promoting the incorporation of flexibility within an intervention strategy. Within this approach, the use of real options is proposed to provide an economic valuation of the flexibility associated with inherently adaptable solutions" (Woodward et al. 2011, 340).

Though there is no specific methodological guidance on how to conduct a real options analysis, Gersonius et al. (2013) summarize a four-step procedure commonly used for ROA in water resources:

1. Specify a scenario tree for a stochastic process.
2. Identify potential options or flexibilities in the system, that is, design variables that can be changed after initial implementation.
3. Formulate the real options optimization problem with regard to objectives, constraints, and decision variables.
4. Run the optimization model.

### Real Options Example

This section presents an example taken directly from HMT and Defra (2009), the United Kingdom's standard procedure for climate change adaptation analysis that emphasizes the real options approaches.

Consider a proposal for investing in infrastructure protecting against the impacts of flooding due to climate change. There are two options: invest in a wall or invest in a wall that can be upgraded in the future, as illustrated in figure 5.2. The simplifying assumptions are that residual damage under "do not invest" strategies have been ignored and the discount factor is 0.8.

## FIGURE 5.2 Illustration of an Upgradeable Wall as an Example of Flood Risk Management Real Options

**50 percent chance high climate impacts**

**50 percent chance low climate impacts**

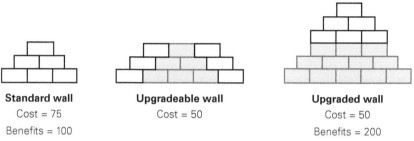

**Standard wall**
Cost = 75
Benefits = 100

**Upgradeable wall**
Cost = 50

**Upgraded wall**
Cost = 50
Benefits = 200

*Source:* Adapted from Woodward et al. 2011.

Assume that there is an equal probability of high and low climate change impacts in the future. The standard wall costs 75, and has benefits of 100 from avoided flooding. The upgradeable wall costs 50, its upgrade costs 50, and the upgraded wall would give benefits of 200 from avoided flooding. The information can be presented in a decision tree, as shown in figure 5.3.

## FIGURE 5.3 Schematic for Real Options Analysis Illustrative Example for Flood Management

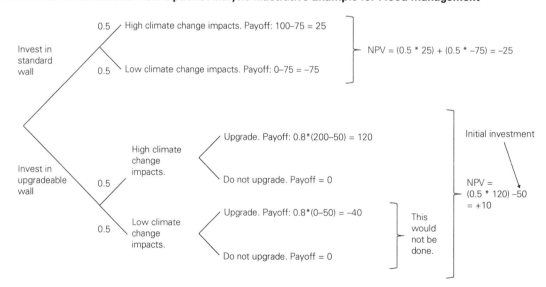

*Simplifying assumptions: Residual damages under "do not invest" strategies have been ignored; the discount factor is 0.8.*

*Source:* Reproduced from HMT and Defra (2009, 15).

*Note:* NPV = net present value.

Confronting Climate Uncertainty in Water Resources Planning and Project Design

The expected value of investing in the standard wall is derived through a simple NPV calculation of the expected costs and benefits of the investment. The NPV is $(0.5 \times 25) + (0.5 \times -75) = -25$. This suggests the investment should not proceed.

Flexibility with regard to the investment decision allows for upgrading in the future if the impacts of climate change are high. The expected value of this option can be calculated. If the impacts of climate change are high enough to warrant upgrading, the value of the investment is 120. If the impacts are low, upgrading is not justified, since the payoff is negative (–40). Because the investment costs of the upgrade are not realized in practice in the low outcome, they are not incorporated into the NPV. The expected value of investing now with the option to upgrade in the future is $(0.5 \times 120) - 50 = +10$.

Comparing the two approaches shows an NPV of –25 for the standard approach, and +10 for the ROA. Flexibility to upgrade in the future is reflected in the higher NPV, and switches the investment decision.

## Summary of Decision Making under Uncertainty

This chapter presents options for use once the water resources program designer (or other user of the decision tree) has deemed climate (and other) information sufficiently well-characterized to proceed to the step of decision making under uncertainty. The options are presented in relation to the phase of the decision tree for which they are most appropriate. This chapter further expands the description of the options for decision making under uncertainty presented in Phase 4 of the decision tree. Though all these tools generally aim at robustness, they differ significantly in their conceptualizations of uncertainty, modeling philosophies, and solution techniques. The tool chosen for decision making under uncertainty should be the tool that best fits the particular water resources context, the project budget and time frame for project evaluation, and the program designer's personal level of comfort with the available methodologies.

## Notes

1. Reoperation is the redevelopment of operation rules for a water system. For example, the operator of a multipurpose reservoir facing a decrease in precipitation might choose to raise the normal maximum water level to better safeguard against drought (at the expense of the capacity of the dam to capture flood flows).

2. Prudhomme et al. (2010) give an example of a well-constructed study that presents future flood risks without providing any method by which planners can decide whether investment in increased robustness is warranted. It can only be assumed that the flood risks presented to the decision maker in this case were somehow weighed against the costs of alternative mitigating strategies, but that the discussion was outside the scope of the publication.

3. Special thanks to David Groves and Robert Lempert for their contributions in the writing of this section about RDM.

4. Special thanks to Jan Kwakkel for his contribution in the writing of this section about DAPP.

## References

Afonso, P. M., and M. d. C. Cunha. 2007. "Robust Optimal Design of Activated Sludge Bioreactors." *Journal of Environmental Engineering* 133 (1): 44–52.

Alcolea, A., P. Renard, G. Mariethoz, and F. Bertone. 2009. "Reducing the Impact of a Desalination Plant Using Stochastic Modeling and Optimization Techniques." *Journal of Hydrology* 365 (3–4): 275–88.

Arrow, K. J., M. Cropper, G. C. Eads, R. W Hahn, L. B. Lave, R. G. Noll, P. R. Portnoy, and others. 1996. *Benefit-Cost Analysis in Environmental, Health, and Safety Regulation.* Washington, DC: AEI Press.

Arrow, K. J., and A. Fisher. 1974. "Environmental Preservation, Uncertainty, and Irreversibility." *Quarterly Journal of Economics* 88 (2): 312–19.

Bau, D. A., and A. S. Mayer. 2006. "Stochastic Management of Pump-and-Treat Strategies Using Surrogate Functions." *Advances in Water Resources* 29 (12): 1901–17.

Bayer, P., C. M. Buerger, and M. Finkel. 2008. "Computationally Efficient Stochastic Optimization Using Multiple Realizations." *Advances in Water Resources* 31 (2): 399–417.

Ben-Haim, Y. 2006. *Info-Gap Decision Theory: Decisions under Severe Uncertainty*, 2nd edition. London: Academic Press.

Ben-Tal, A., and A. Nemirovski. 1998. "Robust Convex Optimization." *Mathematics of Operations Research* 23 (4): 769–805.

———. 1999. "Robust Solutions of Uncertain Linear Programs." *Operations Research Letters* 25 (1): 1–13.

Brown, C., and M. Carriquiry. 2007. "Managing Hydroclimatological Risk to Water Supply with Option Contracts and Reservoir Index Insurance." *Water Resource Research* 43 (11): W11423.

Bureau of Reclamation. 2012. "Colorado River Basin Water Supply and Demand Study: Study Report." United States Bureau of Reclamation, U.S. Department of the Interior, Washington, DC.

Characklis, G. W., B. R. Kirsch, J. Ramsey, K. Dillard, and C. T. Kelley. 2006. "Developing Portfolios of Water Supply Transfers." *Water Resources Research* 42: 1–18.

Charnes, A., and W. W. Cooper. 1959. "Chance-Constrained Programming." *Management Science* 6 (1): 73–79.

Coates, D., D. P. Loucks, J. Aerts, and S. van 't Klooster. 2012. "Working under Uncertainty and Managing Risk." In *Managing Water under Uncertainty and Risk, The United Nations World Water Development Report 4*, Volume 1, edited by Olcay Ünver, 235–58. Paris: United Nations Educational, Scientific and Cultural Organization.

Copeland, T., and V. Antikarov. 2001. *Real Options: A Practitioner's Guide*, 1st ed. New York, NY: Texere.

Cunha, M., and J. Sousa. 2010. "Robust Design of Water Distribution Networks for a Proactive Risk Management." *Journal of Water Resources Planning and Management* 136 (2): 227–36.

de Neufville, R. 2003. "Real Options: Dealing with Uncertainty in Systems Planning and Design." *Integrated Assessment* 4 (1): 26–34.

Dewey, J. 1927. *The Public and Its Problems*. New York: Holt and Company.

Escudero, L. F. 2000. "WARSYP: A Robust Modeling Approach for Water Resources System Planning under Uncertainty." *Annals of Operations Research* 95 (1–4): 313–39.

Friedman, J., and N. Fisher. 1999. "Bump Hunting in High-Dimensional Data." *Statistics and Computing* 9 (2): 123–43.

Gersonius, B., R. Ashley, A. Pathirana, and C. Zevenbergen. 2010. "Managing the Flooding System's Resiliency to Climate Change." *Proceedings of the Institution of Civil Engineers-Engineering Sustainability* 163 (1): 15–22.

———. 2013. "Climate Change Uncertainty: Building Flexibility into Water and Flood Risk Infrastructure." *Climatic Change* 116 (2): 411–23.

Groves, D. G., E. W. Bloom, R. J. Lempert, J. R. Fischbach, J. Nevills, and B. Goshi. 2015. "Developing Key Indicators for Adaptive Water Planning." *Journal of Water Resources Planning and Management* 05014008-05014001 – 05014008-05014010.

Groves, D. G., M. Davis, R. Wilkinson, and R. Lempert. 2008. "Planning for Climate Change in the Inland Empire: Southern California." *Water Resources IMPACT* 10 (4): 14–17.

Groves, D. G., J. R. Fischbach, E. Bloom, D. Knopman, and R. Keefe. 2013. *Adapting to a Changing Colorado River: Making Future Water Deliveries More Reliable through Robust Management Strategies*. Santa Monica, CA: RAND Corporation.

Haasnoot, M., J. H. Kwakkel, W. E. Walker, and J. ter Maat. 2013. "Dynamic Adaptive Policy Pathways: A Method for Crafting Robust Decisions for a Deeply Uncertain World." *Global Environmental Change—Human and Policy Dimensions* 23 (2): 485–98.

Haasnoot, M., H. Middelkoop, A. Offermans, E. van Beek, and W. P. A. van Deursen. 2012. "Exploring Pathways for Sustainable Water Management in River Deltas in a Changing Environment." *Climatic Change* 115 (3–4): 795–819.

Haasnoot, M., H. Middelkoop, E. van Beek, and W. P. A. van Deursen. 2011. "A Method to Develop Sustainable Water Management Strategies for an Uncertain Future." *Sustainable Development* 19 (6): 369–81.

Hall, J., and H. Harvey. 2009. "Decision Making under Severe Uncertainties for Flood Risk Management: A Case Study of Info-Gap Robustness Analysis." *Proceedings of the International Conference on Science and Information Technologies for Sustainable Management of Aquatic Ecosystems*, Concepcion, Chile, IWA.

Hall, J. W., R. J. Lempert, K. Keller, A. Hackbarth, C. Mijere, and D. J. McInerney. 2012. "Robust Climate Policies under Uncertainty: A Comparison of Robust Decision Making and Info-Gap Methods." *Risk Analysis* 32: 1657–72.

Hallegatte, S., A. Shah, C. Lempert, C. Brown, and S. Gill. 2012. "Investment Decision Making under Deep Uncertainty: Application to Climate Change." Policy Research Working Paper 6193, World Bank, Washington, DC.

Hansen, K., J. Kaplan, and S. Kroll. 2014. "Valuing Options in Water Markets: A Laboratory Investigation." *Environmental and Resource Economics* 57 (1): 59–80.

Henry, C. 1974. "Investment Decisions under Uncertainty: The 'Irreversibility Effect'." *American Economic Review* 64 (6): 1006–12.

Hillier, F. S., and G. J. Lieberman. 2005. *Introduction to Operations Research*, 8th ed. Boston, MA: McGraw-Hill.

Hine, D., and J. W. Hall. 2010. "Information Gap Analysis of Flood Model Uncertainties and Regional Frequency Analysis." *Water Resources Research* 46 (1): W01514.

Hipel, K., and Y. Ben-Haim. 1999. "Decision Making in an Uncertain World: Information-Gap Modeling in Water Resources Management." *IEEE Transactions on Systems, Man, and Cybernetics Part C—Applications and Reviews* 29 (4): 506–17.

HMT and Defra (HM Treasury and the Department for Environment, Food and Rural Affairs). 2009. *Accounting for the Effects of Climate Change: Supplementary Green Book Guidance.* London

Ingham, A., J. Ma, and A. Ulph. 2007. "Climate Change, Mitigation and Adaptation with Uncertainty and Learning." *Energy Policy* 35 (11): 5354–69.

Jeuken, A., and T. Reeder. 2011. "Short-Term Decision Making and Long-Term Strategies: How to Adapt to Uncertain Climate Change." *Water Governance* 1: 29–35.

Jeuland, M., and D. Whittington. 2014. "Water Resources Planning under Climate Change: Assessing the Robustness of Real Options for the Blue Nile." *Water Resources Research* 50 (3): 2086–107.

Karmakar, S., and P. P. Mujumdar. 2006. "Grey Fuzzy Optimization Model for Water Quality Management of a River System." *Advances in Water Resources* 29 (7): 1088–105.

Kasprzyk, J. R., S. Nataraj, P. M. Reed, and R. J. Lempert. 2013. "Many Objective Robust Decision Making for Complex Environmental Systems Undergoing Change." *Environmental Modelling and Software* 42 (April): 55–71.

Kasprzyk, J. R., P. M. Reed, B. R. Kirsch, and G. W. Characklis. 2009. "Managing Population and Drought Risks Using Many-Objective Water Portfolio Planning under Uncertainty." *Water Resources Research* 45 (12): 1–18.

Kawachi, T., and S. Maeda. 2004. "Optimal Management of Waste Loading into a River System with Nonpoint Source Pollutants." *Proceedings of the Japan Academy Series B—Physical and Biological Sciences* 80 (8): 392–98.

Kirsch, B. R., G. W. Characklis, K. E. M. Dillard, and C. T. Kelley. 2009. "More Efficient Optimization of Long-Term Water Supply Portfolios." *Water Resources Research* 45 (3): W03414–W03414.

Knight, F. H. 1921. *Risk, Uncertainty, and Profit.* Boston, MA: Houghton Mifflin Company.

Ko, N., and K. Lee. 2009. "Convergence of Deterministic and Stochastic Approaches in Optimal Remediation Design of a Contaminated Aquifer." *Stochastic Environmental Research and Risk Assessment* 23 (3): 309–18.

Korteling, B., S. Dessai, and Z. Kapelan. 2013. "Using Information-Gap Decision Theory for Water Resources Planning Under Severe Uncertainty." *Water Resources Management* 27 (4): 1149–72.

Kwadijk, J. C. J., M. Haasnoot, J. P. M. Mulder, M. M. C. Hoogvliet, A. B. M. Jeuken, R. A. A. van der Krogt, N. G. C. van Oostrom, and others. 2010. "Using Adaptation Tipping Points to Prepare for Climate Change and Sea Level Rise: A Case Study in the Netherlands." *Wiley Interdisciplinary Reviews: Climate Change* 1 (5): 729–40.

Kwakkel, J. H., W. E. Walker, and V. A. W. J. Marchau. 2010. "Adaptive Airport Strategic Planning." *European Journal of Transportation and Infrastructure Research* 10 (3): 227–50.

Lawrence, J., and M. Manning. 2012. "Developing Adaptive Risk Management for Our Changing Climate: A Report of Workshop Outcomes under an Envirolink Grant." The New Zealand Climate Change Research Institute, Victoria University of Wellington.

Lempert, R. J., and M. Collins. 2007. "Managing the Risk of Uncertain Threshold Responses: Comparison of Robust, Optimum, and Precautionary Approaches." *Risk Analysis* 27 (4): 1009–26.

Lempert, R. J., and D. G. Groves. 2010. "Identifying and Evaluating Robust Adaptive Policy Responses to Climate Change for Water Management Agencies in the American West." *Technological Forecasting and Social Change* 77 (6): 960–74.

Lempert, R. J., D. G. Groves, S. W. Popper, and S. C. Bankes. 2006. "A General, Analytic Method for Generating Robust Strategies and Narrative Scenarios." *Management Science* 52 (4): 514–28.

Lempert, R. J., S. W. Popper, and S. C. Bankes. 2002. "Confronting Surprise." *Social Science Computer Review* 20 (4): 420–40.

Li, Y. P., G. H. Huang, Y. F. Huang, and H. D. Zhou. 2009. "A Multistage Fuzzy-Stochastic Programming Model for Supporting Sustainable Water-Resources Allocation and Management." *Environmental Modelling and Software* 24 (7): 786–97.

Li, Y. P., G. H. Huang, and S. L. Nie. 2007. "Mixed Interval-Fuzzy Two-Stage Integer Programming and Its Application to Flood-Diversion Planning." *Engineering Optimization* 39 (2): 163–83.

Loucks, D. P. 1970. "Some Comments on Linear Decision Rules and Chance Constraints." *Water Resources Research* 6 (2): 668–71.

——, J. R. Stedinger, and D. A. Haith. 1981. *Water Resource Systems Planning and Analysis.* Englewood Cliffs, NJ: Prentice Hall.

Loucks, D. P., and E. Van Beek. 2005. *Water Resources Systems Planning and Management: An Introduction to Methods, Models and Applications*. Paris: UNESCO Publishing.

Lund, J. R., and M. Israel. 1995. "Optimization of Transfers in Urban Water Supply Planning." *Journal of Water Resources Planning and Management* 121 (1): 41–48.

Manning, L. J., J. W. Hall, H. J. Fowler, C. G. Kilsby, and C. Tebaldi. 2009. "Using Probabilistic Climate Change Information from a Multimodel Ensemble for Water Resources Assessment." *Water Resources Research* 45 (11): W11411.

Maqsood, M., G. Huang, and J. Yeomans. 2005. "An Interval-Parameter Fuzzy Two-Stage Stochastic Program for Water Resources Management under Uncertainty." *European Journal of Operational Research* 167 (1): 208–25.

Matrosov, E. S., A. M. Woods, and J. J. Harou. 2013. "Robust Decision Making and Info-Gap Decision Theory for Water Resource System Planning." *Journal of Hydrology* 494 (28): 43–58.

McCray, L. E., K. A. Oye, and A. C. Petersen. 2010. "Planned Adaptation in Risk Regulation: An Initial Survey of US Environmental, Health, and Safety Regulation." *Technological Forecasting and Social Change* 77 (6): 951–59.

McDaniels, T., T. Mills, R. Gregory, and D. Ohlson. 2012. "Using Expert Judgments to Explore Robust Alternatives for Forest Management under Climate Change." *Risk Analysis* 32 (12): 2098–112.

Merz, B., J. Hall, M. Disse, and A. Schumann. 2010. "Fluvial Flood Risk Management in a Changing World." *Natural Hazards and Earth System Sciences* 10: 509–27.

Mulvey, J. M., R. J. Vanderbei, and S. A. Zenios. 1995. "Robust Optimization of Large-Scale Systems." *Operations Research* 43 (2): 264–81.

Myers, S. C. 1984. "Finance Theory and Financial Strategy." *Interfaces* 14 (1): 126–37.

National Research Council. 2009. *Informing Decisions in a Changing Climate*. Washington, DC: National Academies Press.

Offermans, A., M. Haasnoot, and P. Valkering. 2011. "A Method to Explore Social Response for Sustainable Water Management Strategies under Changing Conditions." *Sustainable Development* 19 (5): 312–24.

Palmer, R. N., and G. W. Characklis. 2009. "Reducing the Costs of Meeting Regional Water Demand through Risk-Based Transfer Agreements." *Journal of Environmental Management* 90 (5): 1703–14.

Prudhomme, C., R. L. Wilby, S. Crooks, A. L. Kay, and N. S. Reynard. 2010. "Scenario-Neutral Approach to Climate Change Impact Studies: Application to Flood Risk." *Journal of Hydrology* 390 (3–4): 198–209.

Qin, X. S., G. H. Huang, G. M. Zeng, A. Chakma, and Y. F. Huang. 2007. "An Interval-Parameter Fuzzy Nonlinear Optimization Model for Stream Water Quality Management under Uncertainty." *European Journal of Operational Research* 180 (3): 1331–57.

Ranger, N., A. Millner, S. Dietz, S. Fankhauser, A. Lopez, and G. Ruta. 2010. *Adaptation in the UK: A Decision-Making Process*. Grantham Research Institute and Centre for Climate Change Economics and Policy, London.

Ray, P. A., P. H. Kirshen, and D. W. Watkins Jr. 2012. "Staged Climate Change Adaptation Planning for Water Supply in Amman, Jordan." *Journal of Water Resources Planning and Management* 138 (5): 403–11.

Ray, P. A., D. W. Watkins Jr., R. M. Vogel, and P. H. Kirshen. 2014. "A Performance-Based Evaluation of an Improved Robust Optimization Formulation." *Journal of Water Resources Planning and Management* 140 (6). doi:10.1061/(ASCE) WR.1943-5452.0000389.

Reeder, T., and N. Ranger 2011. "How Do You Adapt in an Uncertain World? Lessons from the Thames Estuary 2100 Project." World Resources Report, Washington, DC.

Revelle, C. S., E. E. Whitlatch, and J. R. Wright. 2004. *Civil and Environmental Systems Engineering*, second edition. Upper Saddle River, NJ: Prentice Hall.

Ricciardi, K. L., G. F. Pinder, and G. P. Karatzas. 2007. "Efficient Groundwater Remediation System Design Subject to Uncertainty Using Robust Optimization." *Journal of Water Resources Planning and Management* 133: 253–63.

———. 2009. "Efficient Groundwater Remediation System Designs with Flow and Concentration Constraints Subject to Uncertainty." *Journal of Water Resources Planning and Management* 135 (2): 128–37.

Rosenhead, J. 1989. "Robustness Analysis: Keeping Your Options Open." In *Rational Analysis for a Problematic World*, edited by Jonathan Rosenhead, 181–207. Hoboken, NJ: Wiley.

Rosenzweig, C., and W. D. Solecki. 2014. "Hurricane Sandy and Adaptation Pathways in New York: Lessons from a First-Responder City." *Global Environmental Change* 28 (September): 395–408.

———, R. Blake, M. Bowman, C. Faris, V. Gornitz, R. Horton, and others. 2011. "Developing Coastal Adaptation to Climate Change in the New York City Infrastructure-Shed: Process, Approach, Tools, and Strategies." *Climatic Change* 106 (1): 93–127.

Sahinidis, N. V. 2004. "Optimization under Uncertainty: State-of-the-Art and Opportunities." *Computers and Chemical Engineering* 28 (6–7): 971–83.

Saltelli, A., S. Tarantola, and F. Campolongo. 2000. "Sensitivity Analysis as an Ingredient of Modeling." *Statistical Science* 15 (4): 377–95.

Schultz, M. T., K. N. Mitchell, B. K. Harper, and T. S. Bridges. 2010. "Decision Making Under Uncertainty." ERDC TR-10-12, U.S. Army Corps of Engineers, Engineer Research and Development Center.

Sen, S., and J. L. Higle. 1999. "An Introductory Tutorial on Stochastic Linear Programming Models." *Interfaces* 29 (2): 33–61.

Steinschneider, S., and C. Brown. 2012. "Dynamic Reservoir Management with Real-Option Risk Hedging as a Robust Adaptation to Nonstationary Climate." *Water Resources Research* 48 (5): W05524.

Suh, M. H., and T. Y. Lee. 2002. "Robust Optimal Design of Wastewater Reuse Network of Plating Process." *Journal of Chemical Engineering of Japan* 35 (9): 863–73.

Taguchi, G. 1989. *Introduction to Quality Engineering*. Dearborn, MI: American Supplier Institute.

Tung, Y. K. 1986. "Groundwater-Management by Chance-Constrained Model." *Journal of Water Resources Planning and Management* 112 (1): 1–19.

van Rhee, G. 2012. Handreiking Adaptief Deltamanagement. *Leiden, Stratelligence Decision Support in opdracht van staf deltacommissaris.*

Vucetic, D., and S. P. Simonovic. 2011. "Water Resources Decision Making under Uncertainty." Department of Civil and Environmental Engineering, University of Western Ontario, London, Ontario.

Walker, W. E., S. A. Rahman, and J. Cave. 2001. "Adaptive Policies, Policy Analysis, and Policymaking." *European Journal of Operational Research* 128 (2): 282–89.

Wang, W., K. Chau, C. Cheng, and L. Qiu. 2009. "A Comparison of Performance of Several Artificial Intelligence Methods for Forecasting Monthly Discharge Time Series." *Journal of Hydrology* 374 (3–4): 294–306.

Wang, T., and R. de Neufville. 2005. "Real Options 'in' Projects." 9th Options Annual Conference, Paris, France, June 23.

Watkins, D. W., and D. C. McKinney. 1997. "Finding Robust Solutions to Water Resources Problems." *Journal of Water Resources Planning and Management* 123 (1): 49–58.

Weitzman, M. L. 2009. "On Modeling and Interpreting the Economics of Catastrophic Climate Change." *Review of Economics and Statistics* 91 (1): 1–19.

Woods, A. M., E. Matrosov, and J. J. Harou. 2011. "Applying Info-Gap Decision Theory to Water Supply System Planning: Application to the Thames Basin." Computer Control and the Water Industry (CCWI) Conference, Exeter, UK, September.

Woodward, M., B. Gouldby, Z. Kapelan, S. Khu, and I. Townend. 2011. "Real Options in Flood Risk Management Decision Making." *Journal of Flood Risk Management* 4 (4): 339–49.

Woodward, M., Z. Kapelan, and B. Gouldby. 2014. "Adaptive Flood Risk Management under Climate Change Uncertainty Using Real Options and Optimization." *Risk Analysis* 34: 75–92.

World Bank. 2014. "Enhancing the Climate Resilience of Africa's Infrastructure: The Water and Power Sectors." Africa Development Forum series, World Bank, Washington, DC.

**CHAPTER 6**

# Concluding Remarks

No generally accepted methodology for assessing the significance of climate risks relative to all other risks to water resources projects currently exists. This book puts forth a decision support framework in the form of a decision tree to meet this need. Although the conceptual methodology presented in this book is based on the authors' understanding of the best methods currently available for the assessment and management of climate change risks in water system planning, the procedures presented will benefit from further refinement by repeated application to appropriate pilot test studies in varied geographic, economic, and climate conditions, and with a range of water system planning objectives (municipal water supply, irrigation water supply, flood management, hydropower generation, and so on).[1] The need for updating the outlined procedures will surely become clear as the number of applications grows, and as coincident advances are made in climate science.

One obvious need for innovation is the better integration of techniques for climate change risk assessment with techniques for the assessment of nonclimate risks (economic, political, or natural hazards, for instance), as discussed in relation to Phase 2 Initial Assessment. A second is the need for improved representations of probabilities in the calculation of climate risks.

There are many others. It is hoped and expected that advancements of these and other types will improve subsequent versions of the decision tree framework, resulting in clearer guidance to those whose difficult decisions regarding water system investments under great uncertainty will have lasting ramifications.

## Note

1. The World Bank, for example, has initiated application of the decision tree framework of analysis on a pilot project basis to a hydropower project in Nepal and a water supply and irrigation project in Kenya, to be followed by two additional pilots covering other water system planning objectives. The results of these pilots will provide valuable experiences and lessons to improve the decision tree framework.

# APPENDIX A

# Hydrologic Models

## Introduction

Hydrologic models[1] are simplified, conceptual, mathematical representations of the hydrologic cycle. The typical goal of hydrologic modeling is to simulate the processes of the natural hydrologic system to anticipate future behavior of the system and to predict the system's response changes in forcings (for example, climate change, land use change, water withdrawals). The work of hydrologic models is distinct from that of hydraulic models, which focus rather on simulation of flow patterns in pipes, channels, or porous media. Often, hydrologic models are coupled with hydraulic models in the form of routing models to simulate the streamflow at a given point in a catchment.

Hydrologic models are one component of a coupled hydrologic–water system model typically needed to conduct a climate stress test and analyze the vulnerability of a water infrastructure project to climate change. A water system model translates water volumes as produced by a hydrologic model into economic terms by adding the elements of the anthropogenic system. For example, a hydrologic model might be useful in identifying periods and magnitudes of peak flow at a given point along a river. The system model would translate that peak flow into flood stage, and consequently into flood damage. The water system model will usually include elements of control where the economic impact of infrastructure or policy decisions can be explored or optimized. For a flooding river, the decisions might involve the construction of levees of various heights, reclamation of land within the floodplain, or development of reservoir storage capacity.

Water system decision tools have been discussed throughout this book. This appendix provides a very brief introduction to hydrologic models and offers recommendations for the best tools available for various purposes, with an eye to compatibility with a climate stress test.

In choosing a hydrologic model, the questions to ask are the following: (1) What type of output is needed? (2) What is the basin size? (3) What type of modeling speed is needed?

For a small basin, a lumped model should suffice, and could be built and run quickly. Lumped models have no geographic heterogeneity, and represent all features of the watershed at a point. For larger basins, distributed models are preferred. Distributed models are typically gridded, with each grid cell of the watershed carrying unique model parameters (for example, elevation, ground cover, impervious area). In general, distributed models can be conceptualized as a gridded array of lumped submodels, each connected by routing models (flow vectors) to collect streamflow at the point of interest (typically the lowest elevation point, which is the basin outlet). Likewise, two or more distinct lumped models, each developed at its own time for its own purpose can usually be combined into nongridded (somewhat more organic) versions of distributed models with the inclusion of a routing model. When preexisting, nongridded lumped models are combined using a routing model, the result is referred to as a semi-distributed model (not fully distributed, given that they lack the continuity and uniformity of fully distributed models). The question of how to divide a larger catchment into smaller catchments and when to add a routing model is an area of active research and requires expert judgment.

Typically, by "hydrologic" model, what is meant is "rainfall-runoff" model. The most commonly requested output from a hydrologic model is streamflow, though other outputs are possible, such as soil moisture and groundwater seepage. Explorations of groundwater movement are more appropriate in the realm of porous-media hydraulics than hydrology. The most popular software package for modeling groundwater movement is called MODFLOW. A modular hydrologic system might be a piecemeal assemblage of elements of the total hydrologic system, for example, an element for evapotranspiration (for example, Hamon 1961; Hargreaves 1975), one for groundwater movement (for example, MODFLOW), one for soil moisture (for example, variable infiltration capacity, or VIC), and one for streamflow routing (for example, Muskingum [Cunge, 1969], Soil Conservation Service unit hydrograph, Saint-Venant [1871] equation).

In the United States, the most widely used hydrologic models are the Precipitation Runoff Modeling System (PRMS, used by the United States Geological Survey), the Sacramento Model (used by the National Weather Service for its Advanced Hydrological Prediction Service), the Community Land Model (CLM, of the National Center for Atmospheric Research, and used as an element of general circulation models [GCMs]), the University of Washington's VIC Model, the Soil and Water Assessment Tool (SWAT, developed by the United States Department of Agriculture and Texas A&M), the Water Evaluation and Planning system (WEAP, of the Stockholm Environment Institute, the hydrological modeling component of which is called WATBAL, short for Water Balance), TOPMODEL (of Lancaster University), and the simple, pedagogically excellent abcd model.

Some of these models are physically based and inherently distributed (for example, VIC, CLM, PRMS), meaning that they approximately represent actual natural processes including water, energy, and soil interactions. Some of these models are conceptual (for example, the Sacramento model, the abcd model, WEAP), and keep track of the water balance without reproducing the complicated relationships with energy and soil. Some models straddle the line between conceptual and physically based. If the Sacramento model is coupled with Snow-17,[2] it approximately fits into the partially physically based category. SWAT and TOPMODEL could similarly be described as partially physically based, while TOPMODEL is semi-distributed.

The analogy could be made with methods of GCM downscaling. Conceptual models are like statistical downscaling techniques, which are fast, highly efficient, and typically quite accurate. Physically based models fit the dynamic downscaling metaphor: they are slower, more complex, and more difficult to fit to historical observations. However, the output of physically based models is more useful, easier to interpret, and generally more meaningful to the modeler.

In the opinion of this book's authors, once computing power improves to a sufficient level (and efficiency improvements are made in the hydrologic model algorithms), distributed, physically based watershed modeling systems will be preferred for most applications. Until then, conceptual models will be useful in many cases because of their speed, accuracy, and simplicity. Once certain modifications have been made, and given continuing efforts at better representation of true physical processes in combination with the availability of parallel computing power, physically based models hold great utility.

## Variable Infiltration Capacity (VIC) Macroscale Hydrologic Model

**Source:** University of Washington
**Site:** http://www.hydro.washington.edu/Lettenmaier/Models/VIC/
**Lumped/Distributed:** Distributed
**Conceptual/Physically based:** Physically based
**Appropriateness for climate stress test:** Slow, probably needs parallel computing power to run many thousands of times.

**Other comments:** The VIC model is a macro-scale hydrologic model that solves full water and energy balances, originally developed at the University of Washington. It is physically based and distributed, allowing variation in land cover (vegetation type) in the horizontal plane, as well as variation in soil layer characteristics in the vertical plane. It solves the water balance and energy balance at the same time. The output is physically meaningful and comprehensive (soil moisture layer interactions, runoff from sub-areas, and so forth). Many good sources of input data are available from remote sensing and interpolated ground-based readings. VIC (as well as CLM) is part of global climate models, which integrate GCMs with sea ice and land-surface components.

## Sacramento (originally named the Stanford Watershed Model)

**Source:** National Weather Service River Forecast System
**Site:** http://www.nws.noaa.gov/iao/iao_hydroSoftDoc.php
**Lumped/Distributed:** Lumped
**Conceptual/Physically based:** Conceptual
**Appropriateness for climate stress test:** Highly appropriate, fast.

**Other comments:** What is now known as the Sacramento model was first developed in the 1960s as the Stanford Watershed model—the first known computer-based watershed model. The Stanford Watershed model has many branches. Its immediate successor, the Hydrological Simulation Program-Fortran, "HSPF," is the FORTRAN programming language version developed in the 1970s. When it was translated into the C programming language it became the Sacramento model. The Sacramento model is a conceptual lumped-parameter model focused on soil moisture, which it divides into several components (tension water storage for upper layer and free water storage for lower layer). The Sacramento model also supports different vegetation types and different soil layers. It solves the water balance, and can take into account the energy balance if it is coupled with the SNOW-17 module.

## TOPMODEL

**Source:** Lancaster University
**Site:** http://cran.r-project.org/web/packages/topmodel/index.html
**Lumped/Distributed:** Lumped
**Conceptual/Physically based:** Physically based
**Appropriateness for climate stress test:** Reasonably appropriate, passably fast.

**Other comments:** TOPMODEL encompasses a set of programs for rainfall-runoff modeling in single or multiple subcatchments in a semi-distributed way and using gridded elevation data for the catchment area. It is considered a physically based model because its parameters can be, theoretically, measured in situ (Beven and Kirkby 1979; Beven, Schoffield, and Tagg 1984). TOPMODEL 95.02, written in Fortran 77, is suited to catchments with shallow soils and moderate topography, and that do not suffer from excessively long dry periods. TOPMODEL is a variable contributing area conceptual model, in which the major factors affecting runoff generation are the catchment topography and the soil transmissivity, which diminishes with depth. The present model version includes two mechanisms to estimate surface runoff production: infiltration excess and saturation excess (http://s1004.okstate.edu/S1004/Regional-Bulletins/Modeling-Bulletin/TOPMODEL.html).

Lancaster uses the generalized likelihood uncertainty estimation (GLUE) methodology to carry out calibration, sensitivity analysis, and uncertainty estimation based on many thousands of runs. The current version of TOPMODEL provides an option for output of Monte Carlo simulation results for later use with the compatible GLUE package. Three options are available in the program: (1) the Hydrograph Prediction Option, (2) the Sensitivity Analysis Option, and (3) the Monte Carlo Analysis Option. The results file produced will be compatible with the GLUE analysis software package.

## Water Evaluation and Planning/Water Balance (WEAP/WATBAL)

**Source:** Stockholm Environment Institute
**Site:** http://weap21.org/index.asp
**Lumped/Distributed:** Lumped

**Conceptual/Physically based:** Conceptual
**Appropriateness for climate stress test:** Highly appropriate, fast.

**Other comments:** Recently, an integrated approach to water development has emerged that places water supply projects in the context of demand-side management and water quality and ecosystem preservation and protection. WEAP incorporates these values into a practical tool for water resources planning and policy analysis. WEAP places demand-side issues on an equal footing with supply-side topics. WEAP also distinguishes itself by its integrated approach to simulating both the natural and engineered components of water systems. An intuitive geographic information system–based graphical interface provides a simple, yet powerful, means for constructing, viewing, and modifying the configuration. The Stockholm Environment Institute provided primary support for the development of WEAP. The Hydrologic Engineering Center of the U.S. Army Corps of Engineers funded significant enhancements. A number of agencies, including the United Nations, the World Bank, the U.S. Agency for International Development, the U.S. Environmental Protection Agency, the International Water Management Institute, the Water Research Foundation, and the Global Infrastructure Fund of Japan have provided project support. WEAP has been applied in water assessments in several countries, including Brazil, Burkina Faso, China, the Arab Repbulic of Egypt, Germany, Ghana, India, Israel, Kenya, the Republic of Korea, Mexico, Mozambique, Nepal, Oman, South Africa, Sri Lanka, Thailand, and the United States, as well as countries in central Asia.

## abcd Model

**Source:** Harvard University
**Site:** Not available; the Microsoft Excel version can be requested from the authors of this book
**Lumped/Distributed:** Lumped
**Conceptual/Physically based:** Conceptual
**Appropriateness for climate stress test:** Highly appropriate, fast.

**Other comments:** Alley (1984, 1985), Vandewiele, Xu, and Ni-Lar-Win (1992), and Xu and Singh (1998) compared the performance of numerous alternative monthly water balance models and concluded that a three to five parameter model is sufficient to reproduce most of the information in a hydrologic record on a monthly scale. The abcd model as presented by Thomas (1981) and Thomas et al. (1983) was comparable with other water balance models, with the critical added benefit of simplicity (parameter parsimony), and the supplemental added benefit that each of its parameters has a physical interpretation. Note that "monthly" water balance models may be useful at shorter time scales (that is, daily) if the basin is small enough that all important dynamic processes not included in the model occur within the timestep of the model.

The abcd model is a nonlinear watershed model that accepts precipitation and potential evapotranspiration as input, and produces streamflow as output. Internally, the model also represents soil moisture storage, groundwater storage, direct runoff, groundwater outflow to the stream channel, and actual evapotranspiration. The abcd model was originally introduced by Thomas (1981) and Thomas et al. (1983), at Harvard University. Because of its simplicity and the ability of its few parameters to be described in physical terms, the model makes an excellent pedagogic tool. It also can be developed quickly, and runs quickly, making it an attractive basis for a climate stress test. The abcd model is unrelated to, and has a completely different structure from, the linear "abc" model (Fiering 1967).

## Precipitation Runoff Modeling System (PRMS)

**Source:** United States Geological Survey
**Site:** http://wwwbrr.cr.usgs.gov/projects/SW_MoWS/PRMS.html
**Lumped/Distributed:** Distributed
**Conceptual/Physically based:** Physically based
**Appropriateness for climate stress test:** Slow, probably needs parallel computing power to run many thousands of times.

**Other comments:** PRMS was developed to evaluate the response of various combinations of climate and land use on streamflow and general watershed hydrology. The primary objectives are (1) simulation of hydrologic processes including evaporation, transpiration, runoff, infiltration, and interflow, as determined by the energy and water budgets of the plant canopy, snowpack, and soil zone, on the basis of distributed climate information (temperature, precipitation, and solar radiation); (2) simulation of hydrologic water budgets at the watershed scale for temporal scales ranging from days to centuries; (3) integration of PRMS with other models used for natural resource management or with models from other scientific disciplines; and (4) to provide a modular design that allows for selection of alternative hydrologic-process algorithms from the standard PRMS module library (ftp://brrftp.cr.usgs.gov/pub/mows/software/prms/prmsSummary.txt).

## Community Land Model (CLM)

**Source:** A collaborative project between scientists in the Terrestrial Sciences Section of the Climate and Global Dynamics Division at the National Center for Atmospheric Research and the Community Earth System Model Land Model Working Group.
**Site:** http://www.cgd.ucar.edu/tss/clm/
**Lumped/Distributed:** Distributed
**Conceptual/Physically based:** Physically based
**Appropriateness for climate stress test:** Slow, probably needs parallel computing power to run many thousands of times.

**Other comments:** The CLM is the land model for the Community Earth System Model and the Community Atmosphere Model. The model formalizes and quantifies concepts of ecological climatology. Ecological climatology is an interdisciplinary framework used to understand how natural and human changes in vegetation affect climate. It examines the physical, chemical, and biological processes by which terrestrial ecosystems affect and are affected by climate across a variety of spatial and temporal scales. The central theme is that terrestrial ecosystems, through their cycling of energy, water, chemical elements, and trace gases, are important determinants of climate. The model components are biogeophysics, the hydrologic cycle, biogeochemistry, and dynamic vegetation.

The land surface is represented by five primary, subgrid land cover types (glacier, lake, wetland, urban, and vegetated) in each grid cell. The vegetated portion of a grid cell is further divided into patches of plant functional types (PFTs), each with its own leaf and stem area index and canopy height. Each subgrid land cover type and PFT patch is a separate column for energy and water calculations. The current version is CLM4.0

## Soil and Water Assessment Tool (SWAT)

**Source:** USDA Agricultural Research Service and Texas A&M Agrilife Research
**Site:** http://swat.tamu.edu/
**Lumped/Distributed:** Semi-distributed
**Conceptual/Physically based:** Physically based
**Appropriateness for climate stress test:** Reasonable speed, but less than ideal.

**Other comments:** SWAT (Arnold and Allen 1992; Arnold, Williams, and Maidment 1995) operates continuously, on a daily time step, and is designed to predict the impacts of management practices on hydrology, sediment, and water quality on an ungauged watershed. Major model components include weather generation, hydrology, sediment, crop growth, nutrient, and pesticides. Integration with geographic information systems (GIS) was accomplished by Srinivasan and Arnold (1994). SWAT is a small watershed to river basin–scale model for simulating the quality and quantity of surface and ground water, and predicting the environmental impact of land use, land management practices, and climate change. SWAT is widely used in assessing soil erosion prevention and control, non-point-source pollution control, and regional management in watersheds. SWAT has a high-quality user interface and is particularly user friendly.

## Notes

1. This appendix presents a collection of hydrologic models that, in the authors' experience, are the most trusted and most commonly used hydrologic models in the United States. Inclusion of a particular hydrologic model in this section does not constitute an endorsement of the model by the authors, but rather an expression of its relatively common use and citation. There are, of course, many other hydrologic models not included in this appendix, and their omission should be interpreted as nothing other than the authors' lack of familiarity with them.
2. SNOW-17 is a conceptual model. Most of the important physical processes that take place within a snow cover are explicitly included in the model, but only in a simplified form.

## References

Alley, W. M. 1984. "On the Treatment of Evapotranspiration, Soil Moisture Accounting, and Aquifer Recharge in Monthly Water Balance Models." *Water Resources Research* 20 (8): 1137–49.

———. 1985. "Water Balance Models in One-Month-Ahead Stream Flow Forecasting." *Water Resources Research* 21 (4): 597–606.

Arnold, J. G., and P. M. Allen. 1992. "A Comprehensive Surface-Groundwater Flow Model." *Journal of Hydrology* 142 (1–4): 47–69.

Arnold, J. G., J. R. Williams, and D. A. Maidment. 1995. "Continuous-Time Water and Sediment-Routing Model for Large Basins." *Journal of Hydraulic Engineering* 121 (2): 171–83.

Beven, K. J., and M. J. Kirkby. 1979. "A Physically Based Variable Contributing Area Model of Basin Hydrology." *Hydrological Sciences Bulletin* 24 (1): 43–69.

——, N. Schoffield, and A. Tagg. 1984. "Testing a Physically-Based Flood Forecasting Model (TOPMODEL) for Three UK Catchments." *Journal of Hydrology* 69 (1–4): 119–43.

Cunge, J. A. 1969. "On the Subject of a Flood Propagation Computation Method (Muskingum Method)." *Journal of Hydraulic Research* 7 (2): 205–30.

Fiering, M. B. 1967. *Streamflow Synthesis*. Cambridge, MA: Harvard University Press.

Hamon, W. R. 1961. "Estimating Potential Evapotranspiration." *Journal of the Hydraulics Division* 87 (HY3): 107–20.

Hargreaves, G. H. 1975. "Moisture Availability and Crop Production." *Transactions of the American Society of Agricultural Engineers* 18 (5): 980–84

Saint-Venant, A. 1871. Theorie du mouvement non permanent des eaux, avec application aux crues des rivieres et a l'introduction de marees dans leurs lits. Comptes rendus des seances de l'Academie des Sciences.

Srinivasan, R., and J. G. Arnold. 1994. "Integration of a Basin-Scale Water Quality Model with GIS." *Water Resources Bulletin* 30 (3): 453–62.

Thomas, H. A. 1981. "Improved Methods for National Water Assessment." Report, contract WR 15249270. U.S. Water Resources Council, Washington, DC.

——, C. M. Marin, M. J. Brown, and M. B. Fiering. 1983. "Methodology for Water Resource Assessment." Report to the U.S. Geological Survey, Rep. NTIS 84-124163, National Technical Information Service, Springfield, Virginia.

Vandewiele, G. L., Xu, C.-Y., and Ni-Lar-Win. 1992. "Methodology and Comparative Study of Monthly Water Balance Models in Belgium, China and Burma." *Journal of Hydrology* 134 (1–4): 315–47.

Xu, C.-Y., and V. P. Singh. 1998. "A Review on Monthly Water Balance Models for Water Resource Investigations." *Water Resources Management* 12 (1): 31–50.

**APPENDIX B**

# Worksheets and Report Templates

## Phase 1: The Climate Screening Worksheet

### Step 1

Describe the proposed project and its context. Describe the known and poorly understood parameters that affect project performance.

_____

_____

_____

_____

### Step 2

Is this a water infrastructure project?

• What are the stakeholder-defined performance indicators and risk thresholds?
• What is the project's anticipated economic lifetime?

- What discount rate is preferred (for example, social-welfare or finance-equivalent)?
  → For whom are the project's benefits intended (distributed in space and in time)?
- Is this a rehabilitation or nominal expansion of existing infrastructure?
- Does the project involve a water intake?
  → Is it highly dependent on surface water flows, lakes, or reservoirs?
  → Is it groundwater based? Unconfined or deep (fossil)?
- Does the project involve flood protection?
- Is it dependent on irrigation or domestic water demand?

_____

_____

_____

_____

_____

If this is an infrastructure project, it is very likely to be climate sensitive. If you elect to designate a project of one of these (infrastructure) types as climate insensitive, please provide adequate explanation for the decision, such as impermanence.

_____

_____

_____

_____

_____

**Tip 1:** If the infrastructure project is strictly a wastewater infrastructure project, its climate sensitivities may be relatively small. However, if the project is indeed a pure wastewater infrastructure project, pay special attention to any possible effects of sea level rise on wastewater outfalls to the ocean, and the potential effects of climate change on changes in water use practices affecting the quantity of water entering the sewer network. The **Four C's** will be of particular value in this case.

**Tip 2:** If the proposed project involves a water policy adjustment, training session, environmental or water resources study (without infrastructure), or hydrometeorological service project, the project may be insensitive to climate change. If the proposed project is of this type, and therefore designated as climate insensitive, indicate that here. If the proposed project, *despite being of this type*, may potentially be sensitive to climate change, conduct a **Four C's** analysis.

**Tip 3:** For additional screening guidance, consult the World Bank Climate and Disaster Risk Screening Tools (https://climatescreeningtools.worldbank.org/).

**Tip 4:** For quick identification of anticipated climate changes in the location of the planned project, consult the following:

- The World Bank's Climate Change Knowledge Portal (http://sdwebx.worldbank.org/climateportal/index.cfm)
- The Nature Conservancy's Climate Wizard (http://www.climatewizard.org/).

For geographically targeted resources for climate-change adaptation, consult the United Nations Development Programme's (UNDP's) Adaptation Learning Mechanism (http://www.adaptationlearning.net/).

### The Four C's: Choices, Consequences, Connections, unCertainties

Within the contextual guidance provided by the Four C's, strategic questions are suggested, and the project designer is encouraged to add or delete questions based on relevance to the particular project.

**Choices:** What are the design options? Are different system sizes and configurations possible? Are there substitutable parts? Is the project flexible in time or space? Can the project be modular? What is the timing of the project? Can the project be delayed, or must it be initiated immediately? What are the financing options? Is there wide agreement on the choices available?

_____

_____

_____

_____

**Consequences:** What are the project's benefits and costs, or are other performance indicators primary? How are benefits and costs distributed spatially, temporally, and socioeconomically? How are the project's benefits and costs measured? Can all benefits and costs be monetized? What are the performance thresholds, that is, the criteria for designation as failure or success? What are the windfall possibilities (best case), and what are the worst-case consequences of failure? How might failures be hedged against? Is it possible to assign likelihoods on future scenarios? How might those be quantified and analyzed? Is there wide agreement on the consequences of the project?

_____

_____

_____

_____

**Connections:** How are benefits and costs distributed spatially, temporally, and socioeconomically? How are culture, politics, and environment (among others) affected by this project? Are those effects included in the consequences of the project? Is the success of the project tied to the success of other projects outside the scope of the immediate evaluation? What are the modeling interconnections (that is, what model input must be generated using other models, such as climate model data into hydrologic models, and hydrologic model data into systems models)? What are the scheduling interconnections (that is, what other projects must be completed before this project can begin, and what other projects must wait for this project to be completed)? Is there wide agreement on the project's interconnections?

_____

_____

_____

_____

**unCertainties:** What are the principal uncertainties (1) in the short term (for example, climate, demographic, economic, and political factors) and (2) in the long term (that is, after the first 10–20 years)? Are the uncertainties the result of lack of data or lack of observations, or are they largely irreducible? If the uncertainties can be reduced, how much effort and budget would be required to reduce these uncertainties? Can decisions on the project be delayed until more information is gathered? Is data collection obstructed by matters of national security or secrecy of another kind? How do the uncertainties interact? How might the various uncertainties (principal and minor) be weighted? Are the uncertainties fairly well understood and quantifiable, or would they be better described as "deep"? If trends in available data are discernible, what do the trends indicate, and how well do the historical trends match the projections? Do the various stakeholders agree on the relative significance of the uncertainties? Can a probability distribution be fitted to the uncertainties? Can they be modeled statistically?

_____

_____

_____

_____

## Synthesis

If answers to the Four C's suggest that the project has climate sensitivities, but that those sensitivities are small relative to sensitivities to uncertain factors of other types (for example, demographic or political factors), standard means for evaluating the project (that is, traditional decision analysis) are expected to be sufficient. If, however, the answers to the Four C's suggest that the project's climate sensitivities might be relatively significant, more in-depth quantitative exploration in Phase 2 is warranted.

_____

_____

_____

_____

# Phase 2: Guidance for the Climate Risk Statement

## Introduction

This process is especially relevant to proposed projects shown in Phase 1 to have climate sensitivities, but with questions remaining about the significance of those sensitivities relative to vulnerabilities of other kinds. The Initial Analysis should be used to evaluate the relative significance of climate change vulnerabilities.

Phase 1 projects that are likely to exit the decision tree after a Phase 2 evaluation are wastewater projects; other infrastructure projects with design lives shorter than 20 years; and continuous or long-lasting policies, training activities, or hydrometeorological services. However, hydrometeorological services may have significant sensitivities to climate change, depending on the specific context.

Generally, infrastructure projects with design lives longer than 10–20 years, especially projects in geographic regions with high inter- or intra-annual climate variability, will require more thorough project scoping in Phase 3 (more carefully constructed hydrologic and water resources system models), with particular attention given to capturing potential changes or shifts in climate other than percentage changes in annual average temperature and precipitation. Depending on the results of the Initial Analysis, a number of these projects may require more in-depth climate risk assessment in Phase 3 of the decision tree.

## Rapid Project Scoping

The rapid project scoping procedure described in chapter 3 is one option for execution of a Phase 2 Initial Analysis. Other options are the patient rule induction method (PRIM), or more conventional single-factor sensitivity analysis, for example. Rapid project scoping is based on a simplified analysis of the hydrology and climate change projections (for example, Grijsen 2014) and the evaluation of the elasticity of system performance with respect to climate and other, nonclimate uncertainties. Details of the procedure may be found in "Description of Phase 2" in chapter 3.

## Climate-Related Data Resources

**Tip:** For quick identification of anticipated climate changes in the location of the planned project, consult the following:

- The World Bank's Climate Change Knowledge Portal (http://sdwebx.worldbank.org/climateportal/index.cfm).
- The Nature Conservancy's Climate Wizard (http://www.climatewizard.org/).

For geographically-targeted resources for climate-change adaptation, consult the UNDP's Adaptation Learning Mechanism (http://www.adaptationlearning.net/).

## Synthesis

If the project scoping phase confirms that, though the project has climate sensitivities, those sensitivities are small relative to sensitivities to uncertain factors of other types (for example, demographic or political factors), standard means for evaluating the project (that is, traditional decision analysis) are expected to be sufficient. For expensive or complex projects, it is recommended, given the potential sensitivities identified, that the project designer employ some measure of robustness to the traditional decision analysis, such as safety margins, sensitivity analysis, or adaptive management (see "Extensions to Traditional Decision Analysis" in chapter 5). If climate change–related uncertainties are shown to be relatively significant, more in-depth exploration of those potential vulnerabilities in Phase 3 is warranted.

## Phase 3: Guidance for the Climate Risk Report

Design, execution, and interpretation of the climate stress test require specialized staff or outside expertise.

### Stress Test Procedure

1. At this point, if a formal model of the natural, engineered, or socioeconomic system is not available, it must be created so that climate conditions can be related to the impacts to performance indicators identified in Phase 2.
2. Performance measures and thresholds are established.
3. With the help of expert consultants or highly qualified internal specialists, a standard climate response map is generated for the base design to exhaustively explore the vulnerabilities of the project to change (in climate and other relevant factors). This step is completed using a weather generator and a wide range of projections of uncertainties of other kinds (for example, demographic, economic, land use).
4. Hazards to the project (and their magnitudes) are noted.
5. Probabilities are assigned to subsets of the projected future domain in which particular vulnerabilities are apparent (ex post scenarios). General circulation models first enter the analysis at this point, and are used to inform probabilities in combination with historical trends, paleoclimatology data, and all other useful sources of climate information.
6. Risks are quantified as a function of impacts (hazards) and probabilities.

### The Climate Risk Report

1. Summarizes the procedure used to generate the climate response map, and all assumptions and mathematical relationships.
2. Fully describes all statistical, hydrologic, and water system models (or models of other kinds) constructed for or applied to the climate stress test. The weather generator should receive special attention here.

3. Describes the handling of climate variability (for example, long-term, low-frequency oscillations, as opposed to seasonal oscillations) and any teleconnections to sea surface temperature or other global weather patterns, as well as spatial correlations, temporal correlations, internal variability, and projected mean states (with trends relative to historical trends).
4. Adequately describes data sources (historical observations, general circulation models, paleoclimatology data, and so forth) and computational requirements (for example, number and duration of model runs).
5. In the event that Phase 3 was reentered (and successfully exited) after design modifications were made in Phase 4, the Climate Risk Report should include the Climate Risk Management Plan.

## Phase 4: Guidance for the Climate Risk Management Plan

The Climate Risk Management Plan describes all procedures used to modify the original design. Advanced tools for decision making under uncertainty include robust decision making, information gap decision theory, Dynamic Adaptive Policy Pathways, robust optimization, and real options analysis.

Potential responses to the Phase 4 question: Is Robustness Achievable?

1. If YES, procedures might involve only simple extensions to traditional decision analysis techniques. The Climate Risk Management Plan should describe these procedures, explaining all choices related to performance measures, thresholds, design life, robustness, and flexibility. A rationale should be provided for the nomination of any particular design or set of designs based upon costs, benefits (broadly defined), robustness, and flexibility. The project should then reenter Phase 3, and the Phase 3 Climate Risk Report should summarize the design modifications made and how the new acceptable level of climate risk was achieved. If the level of climate risk identified in Phase 3 for the modified design is not satisfactory, Phase 3 risk assessment and Phase 4 risk management should be iterated until a satisfactory design is achieved.
2. If NO, the project should either be fundamentally revised or abandoned. An abbreviated version of the Climate Risk Management Plan should describe the reasons for abandoning the project. If, alternatively, potentially less climate-vulnerable projects will be proposed, this should be noted.

No in-depth Climate Risk Management Plan is necessary. No Climate Risk Report is necessary.

3. If MAYBE, procedures should probably involve more advanced tools for decision making under uncertainty, such as those described in chapter 5. Procedures involving more advanced tools for decision making under uncertainty may require the services of expert consultants or highly qualified internal specialists.

## Reference

Grijsen, J. 2014. "Understanding the Impact of Climate Change on Hydropower: The Case of Cameroon." Report 87913, Africa Energy Practice, World Bank, Washington, DC.